Upstream Engineering and Flow Assurance Forum 2016

Core Programming Area at the 2016 AIChE Annual Meeting

San Francisco, California, USA
13-18 November 2016

ISBN: 978-1-5108-3451-4

Printed from e-media with permission by:

Curran Associates, Inc.
57 Morehouse Lane
Red Hook, NY 12571

Some format issues inherent in the e-media version may also appear in this print version.

Printed by Curran Associates, Inc. (2017)

For permission requests, please contact AIChE
at the address below.

AIChE
120 Wall Street, FL 23
New York, NY 10005-4020

Phone: (800) 242-4363
Fax: (203) 775-5177

www.aiche.org

Additional copies of this publication are available from:

Curran Associates, Inc.
57 Morehouse Lane
Red Hook, NY 12571 USA
Phone: 845-758-0400
Fax: 845-758-2633
Email: curran@proceedings.com
Web: www.proceedings.com

TABLE OF CONTENTS

(83a) Iron Sulphide Scale Prediction in Oil Wells 1
Giulia Verri, Kenneth. S. Sorbie
(83b) Settling and Re-Entrainment of Wax Particles in Near Gelling Systems 3
Nagu Daraboina, Rama Venkatesan, Cem Sarica, Seetharaman Navaneetha Kannan
(83c) Application of Molecular Characterization to Bituminous Crude Oil to Study Asphaltene Precipitation 4
M. R. Islam, Yifan Hao, Meng Wang, Toni Kirkes, Chau-Chyun Chen
(83d) The Role of n-Alkane Solvent Carbon Number on the Gelation of Long-Chained n-Alkanes in Solution 5
Michael Senra, John Jarboe, Greg Grewal
(83e) Importance of Solubility Parameter Model for Prediction of Asphaltene Precipitation with Regular Solution Theory 6
Nevin Gerek, David Bluck, Seiya Hirohama, Daniel Shaffer, Tuan Trinh
(83g) A Game-Changer on the Modeling Strategy to Predict Asphaltene Precipitation at Reservoir Conditions 7
Mohammad Tavakkoli, Andrew Chen, Francisco Vargas
(173a) Predicting the Viscosity of Characterized Crude Oils 8
Francisco Ramos-Pallares, Helen Lin, Shawn D. Taylor, Marco Satyro, Harvey W. Yarranton
(173b) Investigation of Asphaltene Precipitation from Bitumen As a Function of Diluent Type and Concentration 16
Wattana Chaisoontornyotin, Ng Samson, Michael P. Hoepfner
(173d) A Microfluidic Model for Visualizing Snap-Off during Imbibition 17
Jia Yao, John Oakey
(173e) On the Role of Viscosity Variation on the Onset of Natural Convection 18
Nasser Sabet, Hassan Hassanzadeh, Jalal Abedi
(173g) Solvent Effects on the Structure of Petroleum Asphaltenes 19
Yuan Yang, Michael P. Hoepfner
(173f) Natural Gas Recovery from Kerogen Nanopores 20
Khoa Bui, I. Yucel Akkutlu
(183a) Catalytic Bitumen Partial Upgrading Under Methane Environment 21
Hua Song, Peng He
(183b) Molecular Characterziation and Modeling for Feedstocks Including Heavy Oil and Reactors 22
Shu Wang, Suphat Watanasiri, Darin Campbell
(183c) An Innovative Technique for Oxy-Cracking of Quinolin-65 As a Model Heavy Hydrocarbon Compound 23
Abdallah Manasrah, Nashaat N. Nassar, Amjad El-Qanni, Ismail Badran
(183d) Upgrading of Petroleum Vacuum Residue Using Supercritical Hydrocarbon Solvents in Batch and Continuous Systems 24
Doo-Wook Kim, Chang-Ha Lee
(183e) The Binary Pyrolysis of Sulfur Compounds with Alkyl Aromatics and n-Alkanes 25
Sebnem Hande Ozoren Yasar, Muzaffer Yasar
(183f) Use of Unconventional Resources in State of the Art Oil Refining Processes 26
Refika Cetintas, Ziya Kostereli, Serhat Arca, Elif Kocaman, Emel Baskent, Ramazan O. Caniaz
(183g) Developing a Dead Heavy Oil Viscosity Correlation for Lower Fars, Kuwait 27
Adel Elsharkawy, Saad Alrashdan
(242a) New World of Opportunity Crudes 28
Tim Olsen
(242b) Effect of Nanoporous Monoliths on Saturation Pressure in Shale Reservoirs 29
Hyeyoung Cho, Dominic Caputo, Alberto Martinez, Milind Deo
(242c) Forecasting Gas-Oil Ratios and Solution Gas Production from Liquid Rich Shale Reservoirs 30
Ibukun Makinde, Michael Nikolaou
(242e) Solvent Extraction for Bitumen Recovery from Oil Sands Ores 31
Junyan Wang, Lin He, Hong Sui, Xingang Li
(242g) Short Residue to Useful Products: Different Pathways 32
Utsav Shukla, Utkarsh Maheshwari
(298a) Novel Fracturing Process Utilizing Natural Gas 33
Sandeep Verma, Alhad Phatak, Garud Sridhar, Melissa Poerner, Kevin Hoopes, Griffin Beck

(298b) Comparative Study of Different Enhanced Oil Recovery Methods: A Critical Review 34
Utkarsh Maheshwari, Kaveri Singh
(298c) Modeling of Hybrid Thermal Recovery Methods Using Electromagnetic Heating 35
Asghar Sadeghi, Hassan Hassanzadeh, Thomas G. Harding
(298d) A Journey into the Subnanometer Scale Realm of Organic Matter: An Application of Molecular Dynamics Simulation Approach 36
Gorakh Pawar, Hai Huang
(298f) High Temperature Ultradry Carbon Dioxide-in-Water Foam Stabilized with Viscoelastic Surfactants 37
Keith P. Johnston, Shehab Alzobaidi, Chang Da, Vu Tran
(352a) CFD Simulations and Experiments in Multiphase Flow for Predicting Solid Particle Erosion: Challenges and Success Stories (Invited Talk) 38
Siamack Shirazi
(352b) Development of a Mitigation Strategy for Corrosion-Induced Asphaltene Deposition 39
Mohammad Tavakkoli, Chi-An Sung, Andrew Chen, Francisco Vargas
(352c) A Data-Mining Framework for Uncertainty Analysis in Pipeline Erosion Modeling 40
Selen Cremaschi, Wei Dai
(352d) Impact of Solid Particles and Their Wettability on the Rheological Behavior of Hydrates 43
Ashwin Yegya Raman, Samyukta Koteeswaran, Peter Clark, Clint P. Aichele
(352e) Hydrate Formation in Waxy Oil Systems in the Presence of Surfactants and Solid Stabilizers 44
Deepika Venkataramani, Shane A. Morrissy, Peter Clark, Zachary M. Aman, Clint P. Aichele
(352f) Modeling of Droplet Coalescence and Breakage in Oil-Water Pipeline Using One-Group Interfacial Area Transport 45
Kuochen Tsai
(352h) Flowloop Investigations Towards Gas Hydrate Slurry Transportability 46
Ahmad Abdul Majid, Jose Dapena, Vishal Srivastava, M. Naveed Khan, Michael Volk, Amaka Waturuocha, Ben Bbosa, Carolyn A. Koh
(352i) Hydrate Blockage Risk Assessment in Gas-Dominant Pipelines 47
Thomas Charlton, Zachary M. Aman, Carolyn A. Koh, Michael L. Johns, Eric F. May
(356a) SNG Technology Via Clathrate Hydrates for Large Scale Storage of Natural Gas 48
Hari Prakash Veluswamy, Praveen Linga
(356b) A Water-Immiscible Heavy Guest Molecule for Fast Methane Hydrate Formation 49
Hyery Kang, Yun-Ho Ahn, Dong-Yeun Koh, Seungjun Baek, Jae W. Lee, Huen Lee
(356c) Molecular Dynamics Studies of Structure ?? Hydrate Nucleation Using Advanced Sampling Techniques 50
Ryan Defever, Sapna Sarupria
(356d) Estimation of Direct Transition Mechanism for Molecular Diffusion in Type I Gas Hydrates Using Density Functional Theory 51
Manuel M. Piñeiro, Ángel Vidal-Vidal, Martín Pérez-Rodríguez
(356e) Measurements and Modeling of Gas Hydrate Formation from High Salinity Systems 52
Amadeu K. Sum, Yue Hu, Bo Ram Lee
(356f) Performance of Biodegradable Kinetic Inhibitors on Natural Gas Hydrate Formation 53
Nagu Daraboina, Nicolas Von Solms, Alexander Koltveit
(356g) Towards a Consistent Thermodynamic Theory for Gas-Hydrates 54
Shivanand Vessam, Sudeep Punnathanam
(356h) Exploring the Free Energy Landscape of Gas Hydrate Nucleation By Forward Flux Sampling 55
Yuanfei Bi, Tianshu Li
(451a) Experimental Study of Retention Mechanism of Two Different Types of Scale Inhibitors in Carbonate Formations 56
Khosro Jarrahian, Kenneth. S. Sorbie, Mike. A. Singleton, Lorraine. S. Boak
(451c) Small-Angle Scattering Investigation of the Fractal Structure of Petroleum Asphaltenes 57
Yuan Yang, Wattana Chaisoontornyotin, Michael P. Hoepfner
(451e) A Multicomponent Diffusional Model for Adsorption Dynamics of Asphaltenes at Water/Oil Interfaces 58
Fang Liu, Shaghayegh Darjani, Vincent Pauchard, Nelya Akhmetkhanova
(451d) Comparison of the Vapor-Liquid Equilibria and Solubility Predictions of Various Gas Mixtures Using the Cpa , SAFT and PC-SAFT Equations of State 59
Carolyn A. Koh, Pramod K. W. Harikumar Warrier, Cor Peters, Muhammad Khan
(451f) Amino Acids As Environmentally Friendly Low Dosage Hydrate Inhibitors for Offshore Flowassurance 60
M. Fahed Qureshi, Mert Atilhan, Tausif Altamash

(451g) Inhibited Gas Hydrate Formation at Activated Carbon Particle-Laden Oil-Water Interfaces 61
Seungjun Baek, Juwon Min, Jae W. Lee
(451h) Hydrate Formation in Pickering Emulsions in Flowing Conditions 62
Ashwin Yegya Raman, Deepika Venkataramani, Swanand Bhagwat, Tyler Martin, Clint P. Aichele
Author Index

458633 Iron Sulphide Scale Prediction in Oil Wells

Abstract:

Water flooded oil fields are likely to experience biogenic reservoir souring (in situ H_2S formation) which is the main cause of iron sulphide (FeS) scale problems in this type of production systems.

Despite the increased number of fields around the world affected by this flow assurance problem, the current techniques used for FeS prediction are inaccurate and may lead to inconclusive and misleading results. In fact, FeS scale precipitation depends on in-situ pH, dissolved H_2S, CO_2 and iron concentration which are difficult or in some circumstance impossible to obtain from field samples due to limitations in the sampling and analysis methods.

To overcome this challenge, we have developed a workflow which combines the use of Pressure/Volume/Temperature (PVT) software with Vapour/Liquid Equilibrium (VLE) and scale prediction calculations to model three phase flow rate changes, three phase H_2S and CO_2 compositional changes and pH trends from the reservoir to the first stage of separation using the limited field data available. This information is then used as input for the scale prediction model to provide more reliable scale prediction trends and identify areas in the well with a high scaling risk. When applied to medium/heavy oil wells, there are additional iterative steps required to obtain the three phase H_2S and CO_2 concentrations because we must use a so-called â€œBlack Oilâ€ PVT model which does not accurately account for gas/oil H_2S and CO_2 partitioning.

We applied this new workflow to two different water flooded North Sea oil fields and here we show our results for a representative Well A. This well produces 38°API oil, contains high CO_2 (5% in the gas phase at separator conditions) and low H_2S (500ppmv at the same conditions) but shows FeS precipitation in the produced fluids.

The first step of our workflow is to tune the oil and gas Equation of State (EOS) in the PVT model to obtain the correct GOR and densities. Using this information we can calculate the gas breakout point and understand how the three phase relative volumes change along the well.

Knowing the separator temperature, pressure, gas H_2S and CO_2 mole fraction we can adjust the hydrocarbon composition to obtain the correct concentrations in the gas phase using flash calculations in the PVT model. This will also give us the hydrocarbon H_2S and CO_2 content. Finally, with vapour/liquid equilibrium equations we calculate the water phase H_2S and CO_2 concentration in the separator and using the correct flow rates we determine the total mass of H_2S and CO_2 in the system.

Total H_2S and CO_2 are all recombined to a two phase system in the reservoir where we fix the oil/water partitioning coefficient (form literature values) and calculate the oil and water H_2S and CO_2 concentrations.

When the fluids are taken into the wellbore, the same process is repeated for the two phase system up to the point of gas breakout.

At gas breakout, we run an iterative process to calculate the final H_2S and CO_2 concentration in all three phases which satisfies the mass balance, vapour/liquid equilibrium equation and the fixed oil/water partitioning coefficient. These results give us an H_2S and CO_2 concentration trend from the reservoir to the first stage of separation.

To model the water pH we start from the reservoir using the given water composition, the calculated dissolved H_2S and CO_2 and the carbonate equilibrium ($CaCO_3$ saturation ratio, SR =1) which applies to reservoirs containing carbonate rock.

In the wellbore we take this produced water composition but change temperature, pressure, H_2S and CO_2 to reflect the values associated to each selected point in the well.

Finally, we use the calculated H_2S, CO_2 and pH as data input for the scale prediction model to obtain iron sulphide scaling trends.

The results we obtained show a rather constant pH trend along the well up to the point of gas break out. Due to the high CO_2 concentration, the pH in this region is kept below 6 and the iron sulphide scaling potential slightly decreases as we go up the well thanks to the temperature changes (less precipitation at lower temperature).

However, at gas breakout CO_2 is released and the pH increases causing an increase in iron sulphide scaling potential (which is higher at higher pH).

The pH, bicarbonate and the expected dissolved iron values obtained from our modelling calculations are comparable to field readings and the discrepancies can be explained by the limitations of sampling and analysis techniques as well as uncertainties associated with H_2S solubility in oil and water at low concentrations.

The scaling potential does not change significantly along the well and is limited by the amount of iron available. This information can be used to deploy an effective prevention or mitigation strategy for the control of FeS. The pH values and dissolved H_2S and CO_2 can also be used to run in-situ corrosion calculations and identify areas at higher risk.

Additional challenges associated with FeS scale predictions remain (i.e. prediction of different iron sulphide crystal forms) but this work overcomes the significant implications of pH effect on FeS precipitation.

Applying this newly developed workflow allows any user with access to commercially available PVT and scale prediction tools to answer critical questions related to FeS scaling potential in oil fields which cannot be answered by using these tools individually. This information ultimately allows an operator to deploy effective and efficient mitigation strategies tailored to the sulphide scaling problem.

461402 Settling and Re-Entrainment of Wax Particles in Near Gelling Systems

Abstract:

Wax precipitation and settling is one of the important operational problem in offshore production flow lines during long shutdown especially in near gelling systems. If the settled wax bed does not re-entrain during startup, the fluid is forced to flow in smaller cross-sectional area resulting in several risks, including increased frictional pressure losses, decreased production, and increased risk of equipment failure. Fundamental understanding of the wax settling and re entrainment processes is required to design proper remediation/ prevention strategy. In this work, the settling and re-entrainment phenomena of selected model oil was studied using specially designed transparent flow loop section. Moreover, the important properties of settled wax particles and its influence on re entrainment was investigated.

461934 Application of Molecular Characterization to Bituminous Crude Oil to Study Asphaltene Precipitation

Abstract:

Application of molecular characterization to bituminous crude oil to study asphaltene precipitation

Md Rashedul Islam, Yifan Hao, Meng Wang, Toni Kirkes, and Chau-Chyun Chen

Department of Chemical Engineering, Texas Tech University

Abstract

Bituminous crude oils are vulnerable to asphaltene precipitation when blended with paraffinic samples. Precipitation behavior depends on the nature and content of the paraffins in the blend. A thermodynamic modeling of asphaltene precipitation from crude oil is important as part of the larger effort to understand the compatibility of bitumen or heavy crude oils with the paraffinic ones. Recently, asphaltene precipitations are predicted from wide spectrum of solvents by aggregation thermodynamic framework [1] developed from Yen-Mullins model [2]. However, precipitation prediction from crude oils is still very difficult because of lack of reliable techniques to represent petroleum fluids with real molecules. In this work, a molecule based characterization approach [3] is utilized to represents Cold Lake bitumen with real hydrocarbon molecules. Additionally, asphaltene precipitations are predicted form mixing n-alkanes with Cold Lake bitumen.

References

[1] M. Wang, Y. Hao, M. R. Islam, and C.-C. Chen, "Aggregation thermodynamics for asphaltene precipitation,"

AIChE Journal, vol. 62, pp. 1254-1264, 2016.

[2] O. C. Mullins, "The Modified Yen Modelâ€ ," *Energy & Fuels,* vol. 24, pp. 2179-2207, 2010.

[3] C.-C. Chen and H. Que, "Method of Characterizing Chemical Composition Of Crude Oil For Petroleum Processing," US Patent Application Number: 2013/0185044, July 18,2013.

464706 The Role of n-Alkane Solvent Carbon Number on the Gelation of Long-Chained n-Alkanes in Solution

Abstract:

Crystallization and subsequent gelation of n-alkanes in subsea oil pipelines is an issue of great economic importance to the petroleum industry. To develop better wax remediation strategies, is of interest to gain a better understanding of how n-alkanes crystallize in gel in solution. In this work, the role of the solvent on gelation properties was explored. Model oil solutions containing one or two long chain n-alkanes (carbon numbers of 28, 32 and 36) solutes were dissolved at varying wax fractions in short chain n-alkanes (carbon numbers of 7, 10, 12 and 16). The cloud points, gel points and pour points were measured for these solutions. Results showed that all of the cloud points could be condensed into one relationship using only the mole fraction of the least soluble solute, a result of the solutions all following ideal solution theory. Rheometric results were consistent for samples where the solvent was decane or dodecane. However, hexadecane samples provided results that were much different, an indication that the chain length of the solvent can play a role in the formation of a gel. It is hypothesized that the difference between the trends seen is caused by differences in how the crystals nucleate, grow and aggregate.

465317 Importance of Solubility Parameter Model for Prediction of Asphaltene Precipitation with Regular Solution Theory

Abstract:

Asphaltenes are the heaviest fraction of crude oil and when precipitated produce problems for oil refining. To remove these complex asphaltenes structures, the equilibria of current crude oils and the mechanism of asphaltene precipitation should be evaluated. The regular solution theory has been successfully implemented to predict asphaltene precipitation both for diluted heavy oils and depressurized conventional oils in the presence of a paraffinic solvent [1-3]. The regular solution theory is a thermodynamic approach to solving real solutions that requires molar volume, mole fraction and solubility parameter as input. In this study, we evaluate the effects of input parameters for asphaltene precipitation in n-alkane solvents. Of those, significant effects on the model's accuracy are only observed when varying the solvent solubility parameter and the beta distribution shape factor, while the effect of the number of pseudo-components is negligible beyond a minimum stability value.

Here, we discuss the effect of solubility parameter calculations on the regular solution model. We perform two different models to calculate the solubility parameter of a solvent for the regular solution model of asphaltene precipitation from bitumen mixed with n-alkane solvents. The two methods in question are the thermodynamic (Hildebrand) equation [4] and a linear model [2]. The Hildebrand equation for the solubility parameter produces results that disagree with the correlation between temperature and precipitation yield shown in experimental data. The latter method proposes a linear model for the solubility parameter, however, which follows the same temperature trend as the experimental data. The model is linear with respect to temperature and contains the solubility parameter for the solvent at 25 °C (δ_{25}) as a parameter utilized for the final fitting of the model to the experimental data. The tested solvents for the model are n-heptane and n-pentane, for which the determined δ_{25} values are 7.33 (cal/cm^3)$^{0.5}$ and 6.82 (cal/cm^3)$^{0.5}$, respectively. These solubility parameters maintain validity for various oil samples and temperatures. The drastic effect of the solubility parameter model and its model parameters illustrate the complexity caused by the increased presence of asphaltenes in heavy oil and bitumen samples.

References

[1] H.W. Yarranton, and J.H. Masliyah, "Molar mass distribution and solubility modeling of asphaltenes", AIChE Journal 42 (1996) 3533-3543.

[2] H. Alboudwarej et al., "Regular Solution model for asphaltene precipitation from bitumens and solvents," AIChE Journal 49 (2003) 2948-2956.

[3] K. Akbarzadeh et al., "A generalized regular solution model for asphaltene precipitation from n-alkane diluted heavy oils and bitumens," Fluid Phase Equilibria, 232 (2005) 159-170.

[4] A.F.M. Barton, CRC Handbook of Solubility Parameters and Other Cohesion Parameters, CRC Press, (1991) 24-34.

Acknowledgement: We gratefully appreciate the Schneider Electric internship program.

468687 A Game-Changer on the Modeling Strategy to Predict Asphaltene Precipitation at Reservoir Conditions

Abstract:

The Perturbed Chain version of the Statistical Associating Fluid Theory Equation of State (PC-SAFT EoS) has been successfully applied to model the phase behavior of petroleum systems at reservoir conditions and the prediction of asphaltene precipitation. Although the predictive capabilities of this method and its wide range of applications are much better than other modeling techniques based on conventional cubic equations of state, there are certain features that are still missing, such as the polydisperse nature of asphaltenes. Furthermore, the simulation parameters are usually tuned to Asphaltene Onset Pressure (AOP) data –which are typically obtained by depressurizing live oil samples– or titration experiments of dead oil samples at ambient conditions. These techniques are subject to significant uncertainties caused by the slow kinetics of asphaltene aggregation and the limited sensitivity of the instruments used. The focus of this work is on the experimental study and modeling of asphaltene precipitation from n-alkane - titrated crude oil mixtures considering the polydispersity of asphaltenes. To investigate the effect of asphaltene polydispersity, a crude oil sample is studied by sequential precipitation with different normal alkanes. A recently developed experimental technique called the "Indirect Method" is used for the detection and quantification of asphaltene precipitation. Based on our previous research using the indirect method, it was found that both the onset and the amount of asphaltene precipitation can change with the aging time for a wide range of volume fractions of the added n-alkane. However, the data showed that the amount of precipitated asphaltene at 90 volume percent of a given precipitant is nearly time invariant and therefore this is used to tune the simulation parameters. A distribution function is proposed to represent the asphaltene molecular weight distribution for different asphaltene sub-fractions. It has been found that the inclusion of four asphaltene sub-fractions to represent the polydispersity of asphaltenes is necessary to match the amount of asphaltene precipitated. The modeling results show a reasonable agreement for the precipitated amount at 90 volume percent of n-pentane, n-hexane, n-heptane and n-octane. The PC-SAFT EOS predicts an onset point that requires less addition of n-alkane than the corresponding experimental point after one day of aging time. This result suggests that a thermodynamic limit exists for the onset of asphaltene precipitation and that it corresponds to experiments that require very long aging times, as it has been previously reported by Maqbool et al[1]. By fitting the simulation parameters to the amount of precipitated asphaltenes (and not the detected onset points) at 90 volume percent of added asphaltene precipitant, we can successfully remove the variability of the data with respect to the aging time. This tuning procedure also enables a more accurate prediction of the amount of precipitated asphaltenes at reservoir conditions, which is a more important result for the modeling of asphaltene deposition than the Asphaltene Onset Point itself. The integration of this novel characterization procedure as part of the Asphaltene Deposition Tool (ADEPT)[2–4] developed at Rice University is a topic of ongoing research.

References:

1. Maqbool, T.; Balgoa, A. T.; Fogler, H. S. Energy Fuels 2009, 23, 3681–3686. | **2.** Vargas, F. M.; Creek, J. L.; Chapman, W. G. Energy Fuels 2010, 24, 2294–2299. | **3.** Kurup, A. S.; Vargas, F. M.; Wang, J.; Buckley, J.; Creek, J. L.; Subramani, H., J.; Chapman, W. G. Energy Fuels 2011, 25, 4506–4516. | **4.** Kurup, A. S.; Wang, J.; Subramani, H. J.; Buckley, J.; Creek, J. L.; Chapman, W. G. Energy Fuels 2012, 26, 5702–5710.

PREDICTING THE VISCOSITY OF CHARACTERIZED CRUDE OILS

Francisco Ramos-Pallares. University of Calgary, Calgary, Alberta, Canada
Helen Lin. University of Calgary, Calgary, Alberta, Canada
Shawn D. Taylor. Schlumberger-Doll Research, Cambridge, Mass., USA
Marco A. Satyro. Clarkson University, Potsdam, NY, USA
Harvey W. Yarranton. University of Calgary, Calgary, Alberta, Canada

Introduction

Accurate prediction of the viscosity of whole and diluted crude oils and their fractions is required for process and reservoir simulation. For process simulation, the priority is usually a viscosity model that encompasses the full phase diagram including the critical region. For reservoir simulation, the priority is usually the rapid calculation of gas phase and liquid phase viscosities. While several such models exist for conventional oils, there is a need for a simple yet accurate model for viscosities of mixtures of heavy oil and solvents. In this study, we apply the Expanded Fluid (EF) and the Walther viscosity models to predict the viscosity of whole conventional and heavy crude oils. The EF model is applicable across the phase diagram while the Walther model is intended for single liquid phases. Both have been adapted for crude oils characterized into pseudo-components defined from distillation data.

Viscosity Models

The Expanded Fluid (EF) model (1) relates the viscosity of a fluid to its density as follows:

$$\mu - \mu_G = 0.165\left[\exp(c_2\beta) - 1\right] \tag{1}$$

$$\beta = \frac{1}{\exp\left[\left(\frac{\rho_s^*}{\rho}\right)^{0.65} - 1\right] - 1} \tag{2}$$

$$\rho_s^* = \frac{\rho_s^o}{\exp\left(-c_3|P - P_o|\right)} \tag{3}$$

The inputs of the EF model are the pressure, P in kPa, the dilute gas viscosity, μ_G in mPa.s, and the density of the fluid, ρ in kg/m^3. P_o in Eq. 3 refers to the atmospheric pressure. Additionally, three fluid-specific parameters, c_2, c_3 in kPa^{-1} and ρ_s^o in kg/m^3, are required for any pure component or pseudo-component. EF parameters for over 150 pure hydrocarbons have been determined (2).

The model treats a mixture as one single fluid with fluid-specific parameters calculated from those of the pure components or pseudo-components of the mixture as follows (3):

$$\rho_{s,mix}^o = \left[\sum_i^{nc}\sum_j^{nc}\frac{w_i w_j}{2}\left(\frac{1}{\rho_{s,i}^o} + \frac{1}{\rho_{s,j}^o}\right)\left(1 - \alpha_{ij}\right)\right]^{-1} \tag{4}$$

$$\frac{c_{2,mix}}{\rho^o_{s,mix}} = \sum_i^{nc}\sum_j^{nc}\frac{w_i w_j}{2}\left(\frac{c_{2,i}}{\rho^o_{s,i}} + \frac{c_{2,j}}{\rho^o_{s,j}}\right)(1-\alpha_{ij}) \tag{5}$$

$$c_{3,mix} = \left(\sum_i^{nc}\frac{w_i}{c_{3,i}}\right)^{-1} \tag{6}$$

where nc is the number of components in the mixture, w_i is the mass fraction of the component i, and α_{ij} is a binary interaction parameter. Interaction parameters for EF model are predicted using the specific gravity, SG, and atomic hydrogen-to-carbon ratio, H/C, of the paired components (4) as shown in the appendix.

The Generalized Walther model (5) is given by:

$$\mu = \mu_o\left[1 + \delta(P - P_o)\right] \tag{7}$$

$$\log\left[\log(\mu_o + 1)\right] = A - B\log(T) \tag{8}$$

where μ_o is the viscosity in mPa.s at atmospheric pressure. The inputs of the generalized Walther, or just Walther, model are the absolute temperature, T in K, pressure in kPa and four fluid-specific parameters: a, b, δ and δ_1. The parameters δ and δ_1 are estimated as follows:

$$\delta = \delta_1 - 1.48\times10^{-7}T \tag{9}$$

$$\delta_1 = \frac{0.2715 + 1.4088\ln(MW)}{1\times10^5} \tag{10}$$

The Walther model also treats a mixture as one single fluid with fluid-specific parameters calculated from those of the pure components or pseudo-components as follows (5):

$$A_{mix} = \sum_i^{nc}\sum_j^{nc}\frac{w_i w_j}{2}(A_i + A_j)(1 - \alpha^*_{ij}) \tag{11}$$

$$B_{mix} = \sum_i^{nc}\sum_j^{nc}\frac{w_i w_j}{2}(B_i + B_j)(1 + \alpha^*_{ij}) \tag{12}$$

$$\ln\left[\ln(\delta_{1,mix} + 1)\right] = \sum_i^{nc} w_i \ln\left[\ln(\delta_{1,i} + 1)\right] \tag{13}$$

where α_{ij}* is the Walther model binary interaction parameter and is a function of only the SG of the paired components as shown in the Appendix.

Comparison of Models

The Walther model is only applicable to liquids at reduced temperatures lower than 0.75 based on fitting data for pure hydrocarbons. It requires temperature and pressure inputs and the fitting or prediction of two fluid specific parameters. The EF model encompasses all fluid phases and requires density and pressure inputs and the fitting of two fluid specific inputs. The model is sensitive to small errors in density.

Crude Oil Characterization

A schematic of the characterization procedure for predicting crude oil viscosity is provided in Figure 1and details are reported elsewhere (2). Briefly, the maltene and C5-asphaltene fractions of each oil were characterized separately. The maltenes were split into pseudo-components based on distillation data with a Gaussian extrapolation and the asphaltenes were modeled as a single component.

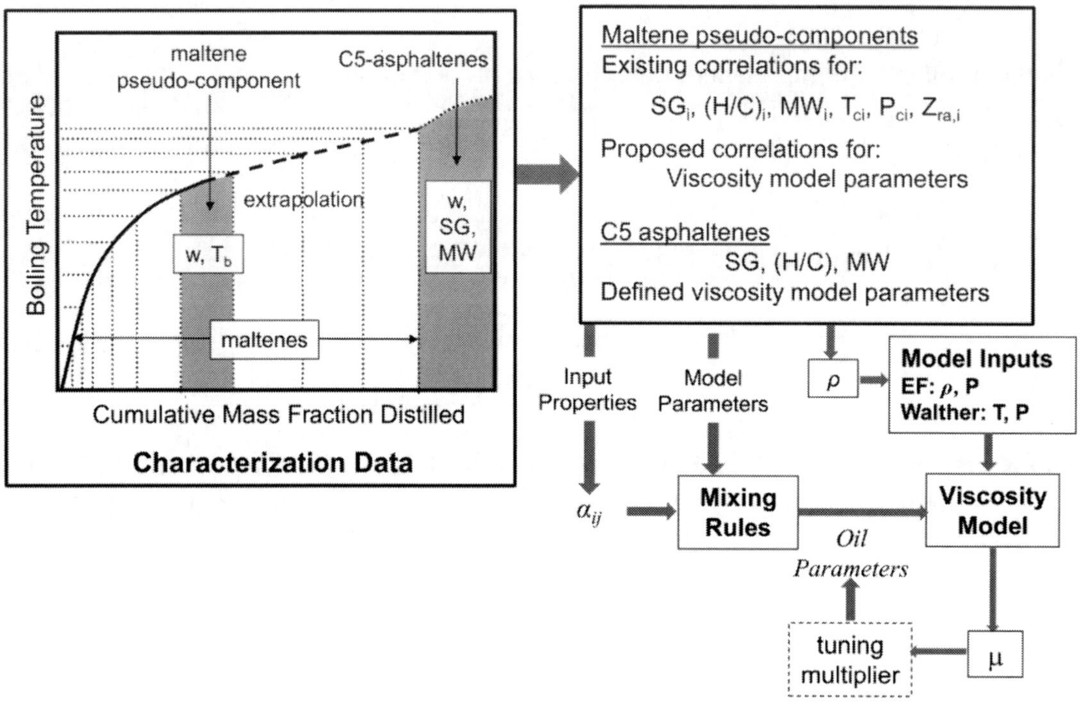

Figure 1. Schematic of characterization procedure for predicting crude oil viscosity from the EF and generalized Walther models.

Maltenes

Normal boiling point, specific gravity, molecular weight, critical properties and hydrogen to carbon ratio are the required properties for viscosity model parameter correlations for maltene pseudo-components. These properties were calculated from a set of existing correlations. The specific gravity of the fraction was calculated from those of the pseudo-components using the following mixing rule:

$$SG = \left(\sum_{i}^{nc} \frac{w_i}{SG_i} \right)^{-1} \tag{14}$$

The calculated maltene specific gravity was tuned to match the experimental value by applying a single-common multiplier to the pseudo-component specific gravities. If the maltene density is not known, the following empirical relation can be used to calculate its specific gravity, SG_{mal}:

$$SG_{mal} = 0.8254\, SG_{oil} + 0.1496 \tag{15}$$

where SG_{oil} is the specific gravity of the oil. The pseudo-component density at 37.7°C is also required as input and is calculated from the Rackett correlation (6). The Rackett compressibility parameter is calculated by matching to the pseudo-component specific gravity. Correlations for the EF and Walther model parameters for maltene pseudo-components are provided in the Appendix.

Asphaltenes
The asphaltene fraction was represented by a single pseudo-component for viscosity modeling purposes with fixed EF and Walther viscosity model parameters determined by fitting each model to Newtonian viscosity data for molten asphaltenes (2). The estimated parameters are reported in the Appendix. Specific gravity and molecular weight are the only physical properties required. The specific gravity and molecular weight were determined indirectly from the measured oil properties, the characterized maltene properties, and the measured mass fraction of C5-asphaltenes in the oil using molar mixing rule and Eq. 14, respectively.

Application to Characterized Oils
To apply the EF and Walther models to predict the viscosity of crude oils, the only inputs required are distillation data, asphaltene mass content and specific gravity and molecular weight of the whole crude oil. Correlations for fluid-specific parameters c_2, ρ_s^o, c_3, A and B are provided in the Appendix. Correlations for parameters δ and δ_l are presented in Eq. 9 and 10, respectively. Details of the calculation procedures for each model are provided below.

EF Model: Parameters for the whole oil, c_2, ρ_s^o, c_3 are calculated from those of the pseudo-components and asphaltenes using the mixing rules (Eq. 4 to 6) with correlated interaction parameters. The density of the oil, required as a model input, is predicted at any temperature and pressure from the pseudo-component and asphaltene densities using Eq. 14. The density of the maltene pseudo-components is calculated as a departure from the atmospheric value using the Tait-COSTALD correlation (7):

$$\rho_{T,P} = \rho_{T,P_o}\left[1 - C\ln\left(\frac{B+P}{B+P_o}\right)\right]^{-1} \tag{16}$$

where $\rho_{T,P}$ is the density of the fluid at temperature, T, and pressure, P, and $\rho_{T,Po}$ is the density at T and atmospheric pressure calculated from the Rackett correlation. C and B are predicted as a function of pseudo-component critical properties using defined correlations (7). The density of asphaltenes is calculated using the following empirical relation assuming that the fraction is incompressible:

$$\rho_{asph} = 1000\,SG_{asph} - \left(6.7424 - 5.098\,SG_{asph}\right)\left(T - 15.6\right) \tag{17}$$

where ρ_{asph} and SG_{asph} are the density and specific gravity of the asphaltenes and T is the temperature in °C. Details are provided in Ref. 2.

Walther Model: Parameters A and B for the whole oil are calculated from those of the maltene pseudo-components and asphaltene fraction using mixing rules (Eq. 11-13) with correlated interaction parameters. Then, the viscosity of the oil at temperature T and atmospheric pressure is calculated using Eq. 8. The viscosity at pressure P is calculated using Eq. 7 with parameter δ calculated from Eq. 9 and 10 using the molecular weight of the oil. Note that the viscosibility parameter δ is estimated from whole oil properties rather than pseudo-component properties.

Comparison of Characterizations: The characterization procedure is almost identical for both models except that a density must be determined at the given temperature and pressure for the EF model but only a specific gravity is required for the Walther model.

Results

Partially Deasphalted Oil and Asphaltene/Toluene Mixture

The estimated asphaltene parameters were tested on a partially deasphalted oil and a solution of 5 wt% C5 asphaltenes in toluene, Figures 3a and 3b, respectively. The samples were modelled as a pseudo-binary mixture of maltenes and asphaltenes or asphaltenes and toluene. The viscosity model parameters for each component were determined by fitting component viscosity data. The viscosity of the oil or solution was calculated using the mixing rules with correlated interaction parameters.

Figure 3a shows that using a single set of either EF or Walther model parameters for the asphaltenes provides satisfactory viscosity predictions even when the asphaltenes are fractionated. It appears that the effect of asphaltenes on the viscosity does not significantly depend on small differences in their molecular weight, self-association, or structural configuration in the crude oil. In addition, asphaltenes in a crude oil and asphaltenes dissolved in a hydrocarbon solvent can be treated in the same way, even though they may self-associate differently, Figure 3b. The viscosity of the partially deasphalted sample was predicted with AARD's of 9 and 12% using the EF and Walther models, respectively. The EF model predicted the viscosity of the asphaltene/toluene mixture with an AARD of 8% whereas the Walther model with an AARD of 12%.

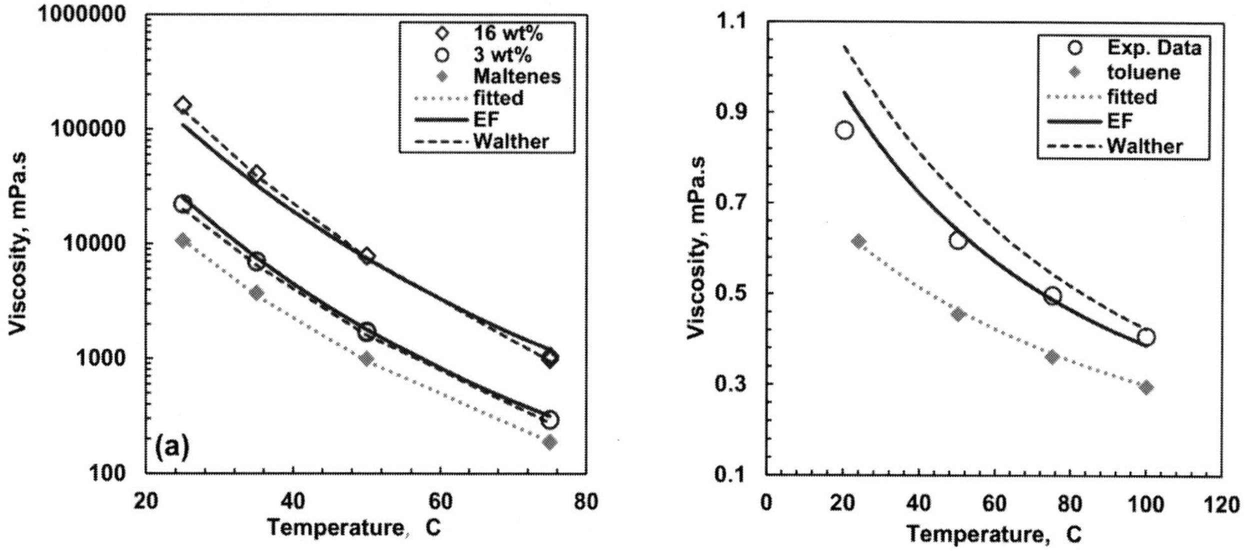

Figure 3. Viscosity versus temperature of: a) partially deasphalted Western Canada Bitumen WC-B-B3 at 0.1 MPa and, b) 5 wt% C5-asphaltenes in toluene at 9 MPa; the asphaltenes were obtained from Western Canada bitumen WC-B-B1. Data taken from Ref. 2.

Predicting Viscosity of Crude Oils

The proposed correlations developed for the EF and Walther model parameters were tested on a dataset of different crude oils from disparate geographical locations. It includes 5 Western Canada and 2 Colombian bitumens; 3 American, 1 Mexican, 1 Venezuelan and 1 Eastern Europe heavy oils and 1 conventional oil from Middle East. The data set includes distillation assay, viscosity and C5-asphaltene content (2). Each crude oil was characterized by splitting into maltenes and asphaltenes as previously described. The maltenes were divided into pseudo-components and the model parameters calculated using the correlations summarized in the Appendix. The asphaltenes were modeled as a single component with defined EF or Walther parameters. The viscosity of the oil was calculated using the mixing rules with correlated interaction parameters. The input density for the EF input was predicted as described previously. Figure 4 shows EF and Walther viscosity predictions for some of the crude oils in the dataset.

Both models are easily tunable in order to improve the accuracy of the predictions as shown in Figure 4b for the Eastern European heavy oil. To do so, at least a single data point at atmospheric pressure must be available. The EF model can be tuned with single multipliers applied to c_2 alone or to both c_2 and ρ_s^o. When both EF parameters are tuned at least two data points must be available.

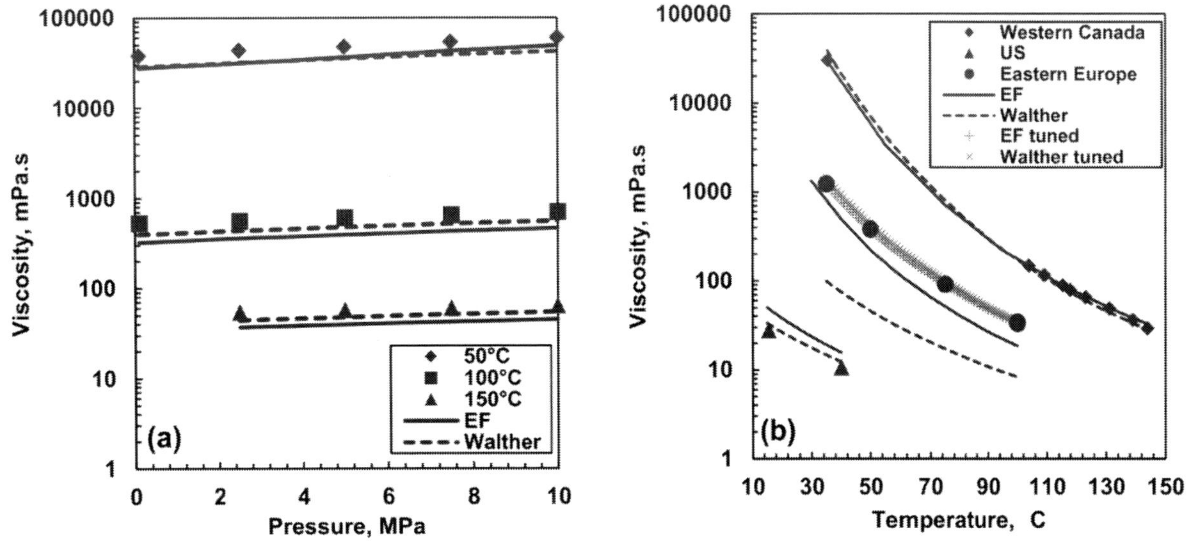

Figure 4. Measured and predicted viscosities of: a) a Western Canada bitumen and, b) another Western Canada bitumen, an Eastern Europe and American heavy oils. EF model was tuned applying a single multiplier to parameter c_2.

Comparison of Results

Both The EF and the Walther models with correlated parameters are capable of predicting the correct trend of viscosity versus temperature and pressure for crude oils of any API gravity. The EF model predicted the viscosity of the all the crude oil samples tested here with AARD of 40% and the Walther model with AARD of 50%. The AARD of the crude oils tested here was reduced to 17% and 7% respectively when c_2 and both parameters were tuned, respectively. In the case of the Walther model a single multiplier applied to the parameter B reduced the AARD of the oils tested here to 8%. There is little difference in the accuracy and tuning of the models except that two tuning parameters were required for the EF model to achieve the accuracy of single parameter tuning with the Walther model.

Conclusion

The EF and Walther models have been successfully used to predict the viscosity of distillation assay characterized crude oils at any temperature and pressure. The methodology is based on a set of correlations developed for maltene pseudo-components and defined model parameters for asphaltenes. EF and Walther model predicted the viscosity of 14 different crude oils with AARD of 40 and 50%, respectively. A simple tuning procedure was proposed to improve the accuracy of calculations. The Walther model is limited to the liquid region. The EF model extends to the entire fluid phase diagram but requires the fluid density as an input.

References

1. Yarranton, H. W.; Satyro, M. A (2009), "Expanded Fluid-Based Viscosity Correlation for Hydrocarbons" *Ind. Eng. Chem. Res.*, 48, pp. 3640-3648.
2. Ramos-Pallares, F.; Taylor, S. D.; Satyro, M. A.; Marriott, R. A.; Yarranton, H. W. (2016), "Prediction of Viscosity for Characterized Oils and Their Fractions Using the Expanded Fluid Model". *Energy and Fuels*, DOI: 10.1021/acs.energyfuels.6b01419.
3. Motahhari, H.; Satyro, M. A.; Yarranton, H. W. (2011), "Predicting the Viscosity of Asymmetric Mixtures with the Expanded Fluid Correlation". *Ind. Eng. Chem. Res.*, 50, pp. 12831-12843.
4. Ramos-Pallares, F.; Schoeggl, F.F.; Taylor, S.D.; Satyro, M.A.; Yarranton, H.W (2016). "Predicting the Viscosity of Hydrocarbon Mixtures and Diluted Heavy Oils Using the Expanded Fluid Model", Energy and Fuels, 30, pp. 3575-3595.
5. Yarranton, H. W.; van Dorp, J.; Verlaan, M.; Lastovka, V. (2013), "Wanted Dead or Live: Crude Cocktail Viscosity-A Pseudo-Component Method to Predict the Viscosity of Dead Oils, Live Oils and Mixtures". *J. Can. Pet. Technol.*, 52, pp. 176-191.
6. Spencer, C. F.; Danner, R. P (1972). "Improved Equation for Prediction of Saturated Liquid Density". *J. Chem. Eng. Data.*, 268, pp. 134-141.
7. Thomson, G. H.; Brobst, K. R.; Hankinson, R. W. (1982). "An Improved Correlation for Densities of Compressed Liquids and Liquid Mixtures". *AIChE J.*, 28, 671-676.

Appendix

Correlations for Maltene Pseudo-Components Parameters

Expanded Fluid Model

$$c_2 = c_2^o - \Delta c_2$$

$$c_2^o = 1.882\times 10^{-3}\exp(0.0058855\,T_b) + 0.3674\,T_b^{-0.1177}$$

$$\Delta c_2 = -2.01417\,\Delta SG^2 - 0.1324\,\Delta SG$$

$$c_3 = \frac{2.8\times 10^{-7}}{1+3.23\exp\left(-1.54\times 10^{-2} MW\right)}$$

$$\rho_s^o = \rho_{37.7^\circ C}\left\{1+\ln\left[1+\frac{c_2}{\ln\left(1+\frac{\mu_{37.7^\circ C}-\mu_G}{0.165}\right)}\right]\right\}^{1/0.65}$$

Generalized Walther Model

$$A = \frac{1000}{9.8071+24.3836\left(\frac{T_b}{T_g^W}\right)}$$

$$\frac{1}{T_g^W} = 0.9482\left(\frac{1}{MW}\right)+0.027$$

$$B = \frac{A-\log\left[\log\left(\mu_{37.7^\circ C}+1\right)\right]}{\log(310.7)}$$

Where MW is the molecular weight, T_b is the normal boiling point in K and the subscript "*37.7°C*" indicates the property at 37.7°C. $\mu_{37.7^\circ C}$ is calculated from the kinematic viscosity and density at the same temperature. The kinematic viscosity, $\nu_{37.7^\circ C}$ in cSt, is predicted according to:

$$\ln\left(\nu_{37.7}+\frac{250}{T_b}\right) = \ln\left(\nu_{37.7}^o+\frac{250}{T_b}\right)\left(\frac{1+2f}{1-2f}\right)^2$$

$$x = 3.7012 - \frac{73.02779}{T_b^{0.5}} \qquad f = -|x|\Delta SG + 53.2315\frac{\Delta SG^2}{T_b^{0.5}}$$

$$\log\left[\log\left(\nu_{37.7}^o+1\right)\right] = (0.0036T_b - 2.0942)0.95^{T_b/200}$$

$$SG^o = 1.098\left[1-\exp\left(-0.00148\,T_b^{1.1128}\right)\right] \qquad \Delta SG = SG^o - SG$$

Asphaltene Parameters

Expanded Fluid Model

$c_2 = 0.9057$, $\rho_s^o = 1113.7\ kg/m^3$

c_3 is calculated from same equation used for maltene pseudo-components.

Generalized Walther Model

$A = 8.3706$, $B = 2.8638$

Correlations for Viscosity Binary Interaction Parameters

Expanded Fluid Model

$\alpha_{ij} = \alpha_{ij}^o - \Delta\alpha_{ij}$, if $\Delta SG_{norm} > 0.165$ $\alpha_{ij}^o = 0.038304 - 0.10478\,\Delta SG_{norm}$

Else $\alpha_{ij}^o = 0.021$. if $\Delta(H/C)_{norm} > 0.25$ $\Delta\alpha_{ij} = 0$

Else $\Delta\alpha_{ij} = 0.02756 - 0.1103\,\Delta(H/C)_{norm}$

Generalized Walther Model

if $\Delta SG_{norm} \le 0.431$, $\alpha_{ij} = 0.025\,\Delta SG_{norm}$

Else $\alpha_{ij} = 0.1318\,\Delta SG_{norm} - 0.04603$

$$\Delta SG_{norm} = \frac{2\left|SG_i - SG_j\right|}{SG_i + SG_j}, \qquad \Delta(H/C)_{norm} = \frac{2\left|(H/C)_i - (H/C)_j\right|}{(H/C)_i + (H/C)_j}, \qquad (H/C)_i = 3.4388 - 1.9327\,SG_i$$

460364 Investigation of Asphaltene Precipitation from Bitumen As a Function of Diluent Type and Concentration

Abstract:

Asphaltene precipitation from diluted bitumen has been extensively studied previously; however, previous work has not considered the slow kinetic precipitation of asphaltenes. Consequently, it is important to revisit asphaltene stability in diluted bitumen and monitor asphaltene kinetic precipitation for more accurate estimates of the asphaltene phase envelope and solubility. The amount of asphaltenes precipitated as a function of diluent type and concentration using various bitumen samples was obtained after monitoring the kinetic aggregation process. Bitumen samples with and without the presence of inorganic solids were evaluated, and naphtha and heptol were used as diluents. The results revealed that the mass of precipitated asphaltenes at equilibrium increases with higher diluent concentrations, in agreement with previous work. At low diluent concentrations, the mass of precipitated asphaltenes was similar regardless of the diluent type. At high diluent concentrations, a large difference was observed in the asphaltene solubility between naphtha and heptol diluents. The results showed that the presence of inorganic solids in the bitumen greatly enhanced the rate of asphaltene precipitation. These results suggest a heterogeneous precipitation mechanism for asphaltenes when inorganic solids are present in bitumen or petroleum samples.

463374 A Microfluidic Model for Visualizing Snap-Off during Imbibition

Abstract:

Oil reservoir rock is typically a porous medium consisting of interconnected microscale pores and throats. The pore-throat geometry of any individual reservoir can vary widely, based upon differing sizes and arrangements of sediment grains. Circular, rectangular and triangular cross-sections are regarded as the most common simplified geometries of pores and throats. Both capillaries with circular cross-sections and microfluidic devices have been utilized as experimental models to study multiphase flow process. However, studies on multiphase flow phenomena in triangular cross-sections are unusual, due to the limitation of triangular cross-section fabrication. To overcome this limitation, we have introduced ultrafast pulsed laser direct writing to quickly produce channels with arbitrary cross section. Femtosecond pulsed laser systems were used to directly write channels in glass with triangular cross-sections. This approach surpasses planar lithography in speed and flexibility, while providing the ability to vary depth within individual channels. We have employed triangular channels of arbitrary cross section to examine the influence of geometry on oil-water snap-off, which is a determining phenomenon in enhanced oil recovery (EOR) processes. In addition, a model with a series of pore-throat arrangements was utilized to show a Rayleigh-Plateau instability induced oil-water or air-water snap-off could be generated solely through the control of the throat geometry, whose cross-section was triangle. This work validates Rayleigh-Plateau theory on snap-off formation in triangular channels and establishes a new class of micromodels for hypothesis-driven EOR studies.

467827 On the Role of Viscosity Variation on the Onset of Natural Convection

Abstract:

Solvent assisted thermal recovery methods have shown promise in improving energy efficiency and reducing environmental footprint of steam based recovery processes. Natural convection may have an important effect during co-injection of solvent and steam into heavy oil reservoirs. It has potential to accelerate the oil recovery due to enhanced mass transfer between solvent and bitumen. In this study, the numerical simulation and linear stability analysis (LSA) for mass and heat transfer during the vertical growth of steam chamber in solvent and steam injection into heavy oil reservoirs are performed. The LSA is conducted based on transient concentration and temperature boundary layers in a porous medium saturated with heavy oil. The bottom surface is subject to a constant solvent concentration and a constant steam temperature. Galerkin method is used to represent the growth of perturbations and the obtained system of ODE equations is integrated using fourth-order Runge-Kutta method.

The obtained scaling relations show that the onset of instabilities is accelerated as the concentration and temperature dependency of heavy oil viscosity is enlarged. Moreover, it is observed that temperature dependency of viscosity has a greater effect on the onset of instabilities than its concentration dependency. However, the presence of solvents is crucial for occurrence of natural convection and they can lead to increased heavy oil production. The results of LSA are compared to the numerical simulation results and good agreement is obtained.

The proposed method provides a more realistic analysis of the problem since it considers transient nature for temperature and concentration fields and also includes the variation of heavy oil viscosity. The introduced scaling relations have a general form and they can be used to predict the onset and wavelength of instabilities under different conditions. Furthermore, the predicted wavelength of instabilities facilitates selection of the proper grid size in numerical simulation of solvent and steam recovery processes.

481134 Solvent Effects on the Structure of Petroleum Asphaltenes

Abstract:

The asphaltene nanoaggregate structure in liquid environments has been recently shown to differ from that of asphaltenes in the solid phase [1]. The correlations between asphaltene molecules can be described by their associated structure, which was shown to change in various solvents. In this study, neutron diffraction has been used to determine the influence of the surrounding liquid on the structure of asphaltene nano-aggregates in mixtures of solvents and anti-solvents (e.g., n-heptane). This talk will present first of their kind neutron diffraction results that were generated for samples containing >5 vol. % asphaltenes dissolved in various deuterated solvents and anti-solvents on the NOMAD instrument at the Spallation Neutron Source at Oak Ridge National Lab. Aromatic, polar and halogenated compounds were used as the solvents, while n-heptane served as the precipitant. Results were instrument, solvent background and inelasticity effect corrected to isolate the scattering signal of the structures formed by asphaltene molecules. This investigation focuses on the molecular spacing between asphaltene structures on the angstrom scale. The separation distance and intensity of each diffraction peak in the system were calculated and compared to previous wide-angle X-ray scattering results [1]. In the neutron diffraction results, the peaks corresponding to the organization of alkyl side chains (Ï' peak) and aromatic core stacking ([002] peak) were observed in agreement with previous X-ray results. The [002] peak is dominated by the correlation between asphaltene structures and represents the crystallinity of asphaltene nano-aggregates. The results showed the crystallinity is constant with changes in the concentrations of asphaltenes or precipitant. A new peak was identified at a separation distance located at a range between 1.7 to 2 angstroms that corresponds to the presence of intermolecular hydrogen bonding between asphaltene molecules. The properties of each diffraction peak vary with different kinds of solvents, which indicate the structure difference caused by solvent effects. The novel results generated with neutron diffraction provide the experimental data as the benchmark for the molecular simulation work and reinforce the need to consider the local solvent matrix on the structure and interactions of asphaltenes.

[1] Hoepfner, M. P. and Fogler, H. S., â€œMultiscale Scattering Investigations of Asphaltene Cluster Breakup, Nanoaggregate Dissociation, and Molecular Ordering,â€ Langmuir, 29 (49), 2013.

455392 Natural Gas Recovery from Kerogen Nanopores

Abstract:

Existing strategies for oil and gas recovery are designed based on macroscopic properties of the produced hydrocarbon fluids. However, recent studies on source rocks revealed that properties of fluids stored in nanopores of organic constituent material kerogen deviate from the bulk behavior. Hence, the traditional equation of state and fluid properties correlations are no longer applicable. This, in turn, leads to added uncertainties in hydrocarbon in-place and recovery calculations for the source rocks that are rich in kerogen. In this paper, we seek to address the question at a fundamental level from the thermodynamics standpoint by simulating isothermal expansion of a quinary hydrocarbon mixture in model nanopore under typical subsurface conditions, and measuring the fluid composition and amount. Molecular Monte Carlo simulations are employed to investigate the equilibrium relationship between the bulk fluid at the outside of the pore and the remaining mixture inside during the stages of pressure depletion. The fluid stored in nanopores shows a composition that varies significantly with the pore size. The smaller the pore is, the heavier its mixture that is in equilibrium with the bulk fluid becomes. During the depletion, the small hydrocarbon molecules escape readily from the pores. The composition of the remaining fluid inside the pore thus becomes progressively heavier and viscous. We show that nanopore confinement limits significantly the release of hydrocarbon molecules from the pores with sizes less than 10nm. For each hydrocarbon component, a strong correlation exists between molar fraction of the component in the produced fluid with that remained inside the pore. This correlation can serve as the basis for establishing alternative methods for the engineering calculations.

472984 Catalytic Bitumen Partial Upgrading Under Methane Environment

Abstract:

Heavy oil and bitumen extracted from oil sands such as the ones abundant in Canada cannot be directly sent to existing refineries mainly due to its high viscosity. Traditionally, hydrotreating is widely employed to break down the long carbon chain and complicated polycyclic molecular structure of heavy crude to lower the viscosity, and remove the impurities such as sulfur, nitrogen and metals simultaneously. However, such treatment consumes expensive hydrogen and requires extremely high operating pressure. Replacing hydrogen with methane, the principal component of natural gas, and lowering the operating pressure would significantly decrease the cost of oil refining. The activation and conversion of methane has been a great challenge due to its symmetric molecular structure and high C-H bond energy. The presence of higher hydrocarbons, which is rich during the upgrading process of heavy oil, would significantly enhance the conversion of methane. In the present work, methane is activated with ZSM-5 based catalysts loaded with active metals and upgrade the heavy oil at low pressure. The charged catalyst witnesses significant viscosity reduction, excellent product stability and compatibility, decreased total acid number, averaged molecular weight, asphaltene content, and increased H/C atomic ratio of the product. The reaction between the heavy oil molecules and methane is studied by choosing several representative model compounds including n-butylbenzene to analog the feedstock components in heavy oil. The study of the reaction is executed from multiple aspects with several analytic instruments, including the GC-MS analysis of products, which provides the composition information, and Diffuse Reflectance Infrared Fourier Transform (DRIFT) Spectroscopy, which detects the formation of organic species on the surface of the catalyst during the upgrading process. ^{13}C and ^{2}D enriched methane are also employed to track the catalytic pathways. Several key products such as ethylbenzene, propylbenzene, isopropylbenzene and butylbenzene, which are the saturation products of styrene, are separated and purified. Their ^{1}H NMR, ^{13}C NMR, FTIR and GC-MS spectra reveal how the CH_x and H are added into unsaturated bonds. This work clearly demonstrates the feasibility of upgrading heavy oil by directly using cheap natural gas on zeolite-supported catalyst at moderate pressure. This

mechanistic study would also benefit the rational design of catalyst for direct upgrading of heavy crude oil using natural gas in the future.

472197 Molecular Characterziation and Modeling for Feedstocks Including Heavy Oil and Reactors

Abstract:

Simulation of refinery-wide processes helps the petroleum industry to design, optimize and operate the refineries profitably and safely, while meeting all necessary regulatory requirements. The traditional pseudo-component approach developed over 30 years ago is still widely used today to represent the feedstocks and products in these process models. This approach uses correlations and heuristics to provide thermodynamic and other properties required in the simulation. For refinery reactor models, kinetic lumps are used to represent molecular classes that undergo reactions to produce the desired products. These kinetic lumps are determined from either finger-print data of the feed or correlations that are based on other properties (such as PNA or sulfur contents) of the feed. Both the pseudo-component and kinetic lump representations do not contain molecular level information. Due to the increasing demands of environment regulations, changing feedstocks, higher margins, etc., the petroleum industry is very interested in improving the accuracy and understanding of the refinery process models and has been exploring the use molecular information for quite some time. In this presentation, we will describe the molecular approach recently developed to leverage molecular knowledge to increase the accuracy of process simulation. The molecular method was applied to characterize crude oils using traditional laboratory assay data (quantitatively match) as well as analytical data such as Fourier Transform Ion Cyclotron Resonance Mass Spectrometry (FT-ICR-MS) (qualitatively match). The resulting molecular compounds, compositions and associated properties are then used to calculate the properties of the process streams. The molecular information are also used directly to represent kinetic lumps composition used in refinery reactor models.

471382 An Innovative Technique for Oxy-Cracking of Quinolin-65 As a Model Heavy Hydrocarbon Compound

Abstract:

Oil refining processes typically involves a coking process that results in a carbon rich solid waste hydrocarbon byproduct called petroleum coke (petcoke). The production rate of the petcoke is over 20 million tonnes per day. Recently, petcoke from Canadian oil sands in U.S. refineries has led to the appearance of large stockpiles adjacent to urban communities in Detroit and Chicago, because limited markets exist for this commodity. The petcoke has strong effects on human and animal health. The main threat to urban populations in the vicinity of petcoke piles is most likely fugitive dust emissions in the form of fine particulate matter. Current industrial processes for treating these kinds of such complicated waste hydrocarbons like direct gasification and combustion have environmental drawbacks centralized mainly in the high production of gaseous CO_2.

In this study, we propose a nano-catalyzed oxy-cracking process that works at mild operating temperature and pressure for high solubilisation of petcoke in relatively low proportions of water. This would open the possibilities for an efficient disposition of petcoke which could be converted at low processing costs into a transportable stream or, through their selective adsorption, be anchored on active catalysts for converting them into lighter products, or alternatively be used efficiently as hydrogen generating source via catalytic steam gasification at much lower temperature than the conventional gasification process.

To understand the mechanism and the kinetics of such innovative technology, Quinolin-65 (Q-65) was selected as a model-oxy-cracking compound mimicking the solid waste hydrocarbons in aqueous-based solutions. In the first phase of this study, the reaction took place in a non-catalyzed batch process operated at optimized oxygen partial pressure and temperature. The preliminary results enabled us to propose a reaction pathway based on the radical reaction mechanism to describe the kinetic behavior of the oxy-cracked Q-65. Hence, kinetic experimental results indicated that the Q-65 oxidation undergoes a parallel consecutive reaction in which an oxidative decomposition took place in the first step producing different aromatic intermediates as confirmed by the NMR analysis. Further, the intermediates were oxy-cracked consecutively into different families of organic acids and small amount of solubilized CO_2 as confirmed by the inorganic carbons analysis. To support our experimental findings, a theoretical computational modeling is conducted. Therefore, the complete radical reaction mechanism will be discussed and elaborated.

465284 Upgrading of Petroleum Vacuum Residue Using Supercritical Hydrocarbon Solvents in Batch and Continuous Systems

Abstract:

As demands for transport fuels and lighter oils have increased, petroleum industries have been forced to refine low-quality extra heavy oils. Among the heavy feedstocks, petroleum vacuum residue (VR), which is the heaviest fraction of crude oil, has high boiling points over 520 °C and contains about 5 wt. % of sulfur, 0.5 wt. % of nitrogen, and organometallic compounds such as nickel and vanadium. Generally, heavy feedstocks can be converted into light products by either carbon rejection process or hydro-conversion process. However, it is difficult for extra heavy oils to be converted into liquid fuels by the conventional upgrading processes due to the high concentration of impurities and asphaltenes.

In the study, supercritical upgrading process (SUP) was applied to upgrade vacuum residue (VR) to light products with small amount of coke formation in batch and continuous systems. Since the physical and transport properties at near-critical region of fluid are altered sensitively, the parametric study for the SUP was required to elucidate optimum operating conditions, depending on various hydrocarbon solvents. The upgrading results were evaluated on residue conversion, coke formation, and product distribution (naphtha, middle distillate, vacuum gas oil, and residue) at various operating conditions; temperature (350-450 °C), hydrogen partial pressure (0-3.45 MPa), reaction time (5-30 min). In addition, further study on a continuous system was conducted on the basis of the results from the batch system. In the continuous system, residence time worked as one of key operating parameters to suppress the coke formation.

470536 The Binary Pyrolysis of Sulfur Compounds with Alkyl Aromatics and n-Alkanes

Abstract:

Petroleum refiners need reliable kinetic models for the optimization of oil residue conversion processes. Modeling of chemical reactions involving heavy ends of petroleum is complicated. Structures and mixtures are very complex, and dissimilar depending on the origin of crudes.

Countless species differ in reactivity should be accounted for in these models, which in turn needs explicit molecular detail of the feed. During conversion many structural changes take place, high boiling compounds broken down into smaller fragments; all sulfur as well as metals such as vanadium and nickel are removed and the most of the aromatics turn into naphthenes which further turn into paraffins and iso-paraffins.

Pure hydrocarbon model compounds pyrolysis gives important information about the reaction kinetics. The purpose of model compounds pyrolysis and its components is to reveal the actual chemical kinetics of the molecules that play a fundamental role during the conversion process. Binary pyrolysis of pure hydrocarbons obtained from the reaction of sulfur compounds reactions will provide important information about heavy oil conversion reactions.

In this study, the binary pyrolysis of decilmercaptane, thiophene n-tetradecane and butyl benzene was investigated at reaction temperatures of of 400-500 °C temperature for reaction time ranging from 30 min to 180 min. In the second phase with the same reaction conditions the binary compounds pyrolysis was carried out. Reaction rate constants and arrhenius parameters for the pyrolysis of binary mixtures obtained from the reaction model were compared with pure compounds.

468016 Use of Unconventional Resources in State of the Art Oil Refining Processes

Abstract:

Conventional oils which are readily accessible and homogenous compared to the unconventional oils are running out. This scarcity draws growing attention to unconventional oils in order to meet the energy requirement of the world. On the other hand, unconventional oils are heavy and complex structures which are tightly trapped between or bound to sand or rock; therefore, new challenges emerge to convert these oils into more valuable products.

In this study, thermal upgrading of oil sand and heavy bitumen derived from domestic reserves is studied to reveal the processibility of these feedstocks in delayed coker units. In addition to that, these feedstocks are blended with highly aromatic refinery by-products to evaluate the effect of aromaticity on coke and liquid yield as well as coke morphology. Feedstock of the delayed coker unit from a refinery is used as a reference for both upgrading and characterization parts.

Oil sands are subjected to solvent extraction method with toluene to separate sand and hydrocarbon samples. After that, these unconventional feedstocks are evaluated for their elemental composition, SARA content, number and weight average molecular weight, specific gravity, dynamic viscosity and S-value (an indicator for stability of asphaltene molecules and coke morphology). Thermal upgrading reactions are conducted at 500 °C for 90 minutes in a tubing bomb reactor having a 10 mL internal volume. Both neat feedstocks and the ones blended with refinery by-products are used in reactions. Coke and liquid products are characterized by means of scanning electron microscopy (SEM) and gas chromatography (GC-SIMDIS), respectively. Moreover, liquid products are further analyzed for their elemental composition, SARA content, molecular weight, specific gravity and dynamic viscosity.

The characterization results indicated that oil sands and heavy bitumen can be used in delayed coker reactor systems but they should be blended with conventional oils to meet design requirements of the units in terms of sulfur and asphaltene content or specific gravity. Furthermore, thermal upgrading reactions demonstrated that unconventional and reference feedstock give nearly same coke yield (~20% wt.) but their liquid product distributions are quite different from each other. Addition of refinery by-products provides around 10% decrease in coke yield which can be attributed to the prevention of phase separation during thermal upgrading reaction by aromatic compounds which can dissolve maltene fractions and asphaltene precursors. Thereby, polycondensation reactions are retarded and coke yield is decreased. Moreover, the S-value results and microstructures from SEM analysis indicated that addition of highly aromatic refinery by-products affects the coke morphology, which specifies the value of coke and occurrence of hot-spot problems in delayed coker units.

Knowing the fact that liquid products are more valuable as compared to solid coke samples whose annual production capacities may reach up to million tons/year for a single refinery, decreasing coke yield and changing coke morphology with easy-to-implement solutions is a powerful tool to increase the profit of oil refineries in a competitive global market.

449940 Developing a Dead Heavy Oil Viscosity Correlation for Lower Fars, Kuwait

Abstract:

Abstract

Viscosity of the produced fluid is an important property for oil and gas industry. It is used in calculations of fluid flow in reservoirs, designing production tubing and flow lines. Accurate estimation of heavy oil viscosity is the most important, as it is one of the key parameters in evaluating a given prospect for the possibility of cold oil production and planning the type of thermal Enhanced Oil Recovery (EOR) methods such as hot water, steam cycling, steam flooding, steam assisted gravity drainage or in-situ combustion. Viscosity of crude oil is usually measured at reservoir temperature at various pressures using bottom-hole sample or surface recombined sample.

Occasionally, the sample is difficult to obtain or becomes unreliable. In this case calculation methods are used to estimate the crude oil viscosity as a function of easily measured production information such as API gravity or detailed compositional data. Many correlations have been presented to estimate crude oil viscosity for crude oils of a certain field or region. Application of these correlations to other parts or regions results in large error in estimated viscosity because each crude oil has its own physico-chemical characteristics that makes it different from other crudes.

In this study, the accuracy of the well-known correlations are tested against heavy oil viscosity data collected from Lower Fars formation in Kuwait. Later, a new correlation is presented to estimate the dead oil viscosity of Lower Fars. Accuracy and reliability of the proposed model is compared to published correlations such as Beal (1946), GlasØ (1980), Kartoatmodjo and Schmidt (1994), and Al-Omair et al. (2014).The results showed that the new correlation is able to predict the dead oil viscosity with mean absolute relative error of 0.16, correlation coefficient equal of 96%, and standard deviation of 0.17.

473688 New World of Opportunity Crudes

Abstract:

The refining industry has changed noticeably over the past several years with an ample supply of opportunity crude oils available, which include light tight oils and heavy sour crudes, and sometimes high TAN crudes. Opportunity crudes have been around for years, but only until recently has the abundance of these discounted crudes shifted the behavior of refiners.

With the use of opportunity crude oils comes new processing challenges; Automation technology is advancing to meet these challenges to ensure refiners are successfully alleviating these new issues. Advances in real time crude analysis and crude unit monitoring and control allow refiners to maximize opportunity crude processing while mitigating issues like accelerated fouling and corrosion.

Refiners on the coast will typically have access to more opportunity crudes than those landlocked. Because crude oils can be very different in properties, refiners try to match the required crude oil composition for their refinery configuration; usually by blending two or more crude oils. To make matters more challenging, sometimes crude oils purchased to process are crudes that the refiner has no experience processing.

Crude blending operation has become more important with the ability to establish consistent crude feed properties. Steady operation enables better optimization of the crude unit by pushing against the appropriate constraints which often includes maximizing the quantity of discounted opportunity crude oils blended.

This presentation will highlight on new operating challenges, and show a couple examples where refiners have made modifications in both technology and behavior to adapt to the new world of opportunity crudes.

463572 Hydrocarbon Mixture Bubble Points in Nanoporous Silica Monoliths

Abstract:

Shale reservoirs are now becoming the focus of new oil and gas exploration and development. These shale reservoirs usually have lower permeability and porosity with a pore size distribution ranging between 1-20 nm. It is hypothesized that fluids in the confined systems (shale reservoirs) have different thermo-dynamic properties than conventional reservoirs. However, because of the difficulty of the realization of the pore structure of shales, there have been few experimental studies. In my research, two different nanoporous silica based monoliths were synthesized with similar pore size ranges as the shale reservoirs, represented in the porous system of shale reservoirs. Monoliths can be synthesized into a desired shape by putting them in a customized container. Because of this advantage, the synthesized monoliths have only nano-sized pores, not macropores. Two nanoporous monoliths were synthesized by modifying the synthesizing procedures that could be used to synthesize hexagonal and cubic mesoporous silica powders, which have different morphologies. From comparing the saturation pressure with and without nanoporous media, we can confirm the hypothesis that nano-sized pores have an effect on saturation pressure. Also, we can see the different morphologies of the two monoliths have an effect on saturation pressure. These synthesized monoliths have been characterized by X-ray diffraction (XRD), nitrogen adsorption/desorption isotherm (BET), and transmission electron microscopy (TEM), which show their physicochemical properties. The phase diagram of methane and decane mixture has been modeled with the Winprop module from the CMG reservoir simulator. The experimentally measured saturation pressure matches with the simulated result. However, the measured saturation pressure changes with nanoporous media.

455711 Forecasting Gas-Oil Ratios and Solution Gas Production from Liquid Rich Shale Reservoirs

Abstract:

The basic issue: To enhance our knowledge about the production mechanisms of shale oil reservoirs and better exploit their economic potential, it is important to be able to forecast production of the secondary phase, namely gas.

Why it is important: Stakeholders often focus on oil forecasts when it comes to shale oil reservoirs, ignoring the equally important solution gas produced from these plays. Lack of production data, complicated flow mechanisms of liquid rich shale reservoirs, production pattern of producing gas-oil ratios among other factors, make the task of forecasting solution gas production difficult.

What we are proposing: The objective of this paper is to present a new method for estimating producing gas-oil ratios (GOR) and solution gas production from shale volatile oil reservoirs. The method is data driven while observing the fundamental equations that govern hydrocarbon production over time. It entails the following steps:

Step 1 – Generate representative collection of data: To develop the functional form that describes the evolution of GOR over time for a producing well, run a large number of representative simulations over a fairly long production time (e.g. 30 years) so that a variety of wells in various reservoirs under diverse producing conditions can be well represented in the sample.

Step 2 – Identify natural GOR shapes through PCA: Stack GOR profiles on top of each other, to construct a matrix of dimension perform principal component analysis (PCA) on , and then approximate as , where . The vectors can be thought of as natural shapes over time, whose linear combination can reproduce – with reasonable approximation – every conceivable GOR profile over time.

Note:Steps 1 and 2 will have to be completed perhaps only once or at worst a limited number of times, since the natural GOR shapes they produce will be universal.

Step 3 – Fit a linear combination of natural GOR shapes to data over limited time: Given GOR data for a limited production history over time , identify estimates by finding .

Step 4 – Use the model of Step 3 for future predictions: The value of GOR at a future time along with a corresponding confidence interval can be estimated using standard formulas as where is the information matrix.

Case study: A commercial compositional simulator was used to simulate 30 years of production from a shale volatile oil reservoir with 4 different reservoir fluids. The reservoir permeability and porosity were 0.001 md and 6% respectively. Several values for the following parameters were considered in a total of the simulations:

1. 1. Critical gas saturations between 5% and 20%;
2. 2. Constant bottomhole pressures (BHP) between 500 and 1500 psi; and
3. 3. Initial reservoir pressures between 3000 and 5000 psi.

PCA on the GOR data produced by these simulations revealed (through cross-validation) that a small number of principal components adequately capture the GOR profiles observed over the 30-year period. To test how well the resulting model can forecast future production from production data over limited time, we considered a horizontal well of length 5000 ft, with 20 hydraulic fractures of 150 ft half-length, spaced 250 ft apart. We simulated producing GOR over 30 years, fit the model to the corresponding simulated data over periods from 6 months to 3 years, and then compared GOR forecasts by the model to GOR values produced by the simulator into the future. The future profiles produced by the PCA model matched those produced by the simulator with reasonable level of accuracy. The PCA model was also found to match the simulator on solution gas production, calculated by integrating producing GOR over cumulative oil production.

Conclusions: This work presents a novel approach of forecasting production from liquid rich shale plays, based on a well established statistical approach. It bypasses the need to invent an ad hoc functional form for GOR over time that would be both simple for computation and realistic enough to capture real data. The proposed approach can help improve reserves estimation, reservoir management, and project economics.

467020 Solvent Extraction for Bitumen Recovery from Oil Sands Ores

Abstract:

Unconventional oils, including oil/tar sands, shale oil, heavy oil, etc., have been considered as an alternative of traditional crude oils due to their huge reserves (about two thirds of the total oils) buried around the world. It is estimated that there are about 1700 trillion barrels of bitumen contained in the oil sands ores in Alberta, Canada(Chew, 2014; Gray et al., 2009). To release this kind of petroleum, hot water-based extraction (HWBE) has been proposed and applied in industry which creates large volumes of wet tailings with fine solids. To avoid this environment pollution, the solvent extraction has been attracting increasing attention from industry due to its high extraction efficiency, less water and energy consumption, and high compatibility for different kinds of ores.

In the present study, solvent extraction of oil sands has been investigated using pure organic solvents, including toluene, tetrahydrofuran (THF), cyclohexane, n-heptane, n-hexane, ethyl acetate, and etc. Results of bottle tests showed that the bitumen recovery increased with the increase of the number of carbon and the cyclic hydrocarbons (toluene, THF and cyclohexane) performed better than alkane hydrocarbons (n-heptane, n-hexane, ethyl acetate and acetone). To analysis the bitumen in the extracted phase more detailedly, it has been divided into four fractions, named as saturates, aromatics, resins, and asphaltenes (SARA). Results show that increasing the aromaticity of the solvent is beneficial for the recovery of heavy fractions (e.g., resins and asphaltenes) of bitumen from the oil sands.

To further explore the solvent extraction process, solubility simulation in COSMOtherm software was applied and four typical molecules were selected as the representatives of SARA. A similar trend of bitumen recovery with different solvents was obtained as an order of: THF â‰¥ toluene > cyclohexane > n-heptane > n-hexane > ethyl acetate. Firstly, the efficiency of solvent extraction is found to be highly dependent on the polarity of the solvent. The solvents with too low (i.e. n-hexane and n-heptane) or too high (i.e. acetone and methanol) polarity had a poor solubility to bitumen. The polarity of the solvent will cause certain effect to the bitumen recovery because some special polar groups (i.e. groups containing oxygen, nitrogen, sulfur, and etc.) affected the properties of bitumen components. Secondly, the enhancement of solvent aromaticity makes the recovery of aromatics, resins and asphaltenes increase. The reason for these may be that the heavy bituminous constituents contain aromatic or cyclic structures. From the solubility point of view, the more similarity of microstructure between bitumen components and solvent, the more bitumen was recovered. Finally, after careful analysis, it is found that the dispersion force of the solvent plays a dominate role (weight > 55%) in dissolving bitumen fractions, followed by hydrogen-bonding force (weight at 0~26%) and polar force (weight at 0~18%). This finding allows us to consider more detailed influence factors when selecting the potential solvents for bitumen recovery process and solubility simulation may be a relatively efficient and accurate previous prediction of solvents screening.

Accordingly, the solvent extraction process would be a promising method for the recovery of unconventional oil from the ores. This study would improve the knowledge base for bitumen recovery mechanism and serve for future work on the solvent extraction technology applications.

References:

Chew, K.J., 2014. The future of oil: unconventional fossil fuels. Philos Trans A Math Phys Eng Sci 372, 20120324.

Gray, M., Xu, Z., Masliyah, J., 2009. Physics in the oil sands of Alberta. Phys. Today 62, 31-35.

464457 Short Residue to Useful Products: Different Pathways

Abstract:

With the depletion of the oil reserves, the oil is getting heavy day by day. The utilization of the heavy oil in the refinery process leads to the generation of high amount of short residue from the vacuum distillation units. The present study is about finding the different pathways for the utilization of the short residue to the formation of useful products. The conversion of short residue to useful product also helps in increasing the economic viability of the refinery processes. This research work initially studies about the quality of the various crudes available in the Indian petroleum industry. The different processes involved for the upgradation are categorized as Carbon Rejection Process and Hydrogen Addition Process. The present work further discussed on the selection of the specific process depending on the quality of the short residue left in the refinery.

474557 Novel Fracturing Process Utilizing Natural Gas

Abstract:

Hydraulic fracturing is used to increase oil and natural gas production by injecting high-pressure fluid (primarily water) into a rock formation to fracture the rock and release trapped oil and natural gas. This method was developed to increase yield and make production feasible in areas that would not be viable using traditional drilling technologies. Hydraulic fracturing does have some drawbacks, however, including excessive water use and the large environmental footprint associated with that water usage. Other fracturing fluids such as carbon dioxide and propane have been tested to try to reduce the environmental impact, but the related processes are expensive and have only been implemented on a limited scale.

In the past year, under a Department of Energy (DOE) program, a novel process has been developed that utilizes natural gas as the fracturing fluid. Using natural gas requires minimal water consumption because the fracturing fluid is composed mostly of produced natural gas from adjoining wells. One challenge associated with using produced natural gas is designing and optimizing a mobile process to produce the flow rate and pressure required for fracturing. As a part of the DOE program, several different thermodynamic cycles were considered for processing the natural gas. Options included direct compression and various refrigeration cycles to produce liquefied natural gas, which was then pumped to high pressure before vaporization. Each of these cycles was evaluated and the optimal cycle was then designed in detail. Some of the factors considered in the selection included specific power, cost, size, weight, and mobility. This presentation reviews the various designs that were considered and outlines the method used to select the optimal process. A description of the final detailed design is presented.

464381 Comparative Study of Different Enhanced Oil Recovery Methods: A Critical Review

Abstract:

With the depletion of the petroleum reservoirs throughout the world, there is an immense utilization of the various enhanced oil recovery techniques. The enhanced oil recovery techniques range from thermal to non-thermal techniques. The thermal methods can be using hot water or steam as well as using electrical heating. The non-thermal methods included the miscible flooding, non-miscible flooding and chemical flooding. The present research deals with the study of various enhanced oil recovery techniques utilized for the extraction of the leftover oil in the reservoirs. It also studies about the economic benefits of the different enhanced oil recovery techniques. The later part of the study deals with the utilization of the properties of the leftover reservoir for the selection of the available enhanced oil recovery techniques.

467836 Modeling of Hybrid Thermal Recovery Methods Using Electromagnetic Heating

Abstract:

Conventional thermal methods such as the cyclic steam stimulation (CSS) and steam assisted gravity drainage (SAGD) have been implemented in the last decade. The main obstacle with these methods is the efficiency of heat delivery to reservoir due to heterogeneous nature of oil-bearing formations. High steam oil ratio, or energy costs per produced barrel of oil, has been always an issue with thermal recovery methods. The second challenge has been environmental footprint of thermal oil recovery methods (greenhouse gas emissions and water usage/treatment). In situ heating of bitumen and heavy oil reservoirs, such as electromagnetic heating (EMH), does not require carrier fluids and is less sensitive to reservoir geology. Therefore, it provides an efficient way of exploitation from difficult reservoirs with less environmental impact and comparable or even less energy intensive operations. Furthermore, the accuracy and control on heating provided by the electromagnetic heating may offer a potential to improve the efficiency of the thermal process and attenuate its environmental drawbacks. However, commercialization and implementation of EMH technique requires comprehensive research and development, pilot tests, and engineering and life cycle analysis before the application to a real field. Unlike conventional heating approaches, which are dominated by convection and conduction, the electromagnetic heating involves variations of the electric and magnetic fields in space and time. Because of computational complexity and computational time limitations Lambert's law has been proposed for high frequency waves. In this study, for the first time the applicability of Lambert's law for bitumen extraction is investigated. The range of electrical properties of bitumen bearing formation at which the Lambert's law meets the fundamental equations of Maxwell have been identified. In addition, analytical models of co- and counter-current high frequency heating for linear systems for a SAGD well configuration has been developed and energy efficiency of high frequency electromagnetic heating is compared with the traditional approach of steam circulation.

472444 A Journey into the Subnanometer Scale Realm of Organic Matter: An Application of Molecular Dynamics Simulation Approach

Abstract:

The subnanometer scale realm of organic matter has remained unexplored due to the limitations of the conventional visualization techniques such as scanning electron microscopy (SEM)/transmission electron microscopy (TEM), which are unable to resolve the underlying length scales. We report a molecular dynamics simulation-based approach to investigate the subnanometer scale kerogen structure and to evaluate the key kerogen attributes â€" subnanometer scale pore size distribution, kerogen porosity, and specific surface area â€" which unavoidably governs the state of hydrocarbons (free vs. adsorbed) trapped in subnanometer kerogen pores and also plays a crucial role in the primary expulsion (of hydrocarbons) from kerogen. The elucidation of subnanometer scale features in organic matter can significantly advance the current understanding of organic matter and the development of new strategies to enhance the hydrocarbon recovery from nanoporous geological formations such as shale reservoirs.

Keywords: Organic matter, Subnanometer scale characterization, Molecular dynamics

478753 High Temperature Ultradry Carbon Dioxide-in-Water Foam Stabilized with Viscoelastic Surfactants

Abstract:

Ultradry carbon dioxide- in-water (C/W) foams with gas phase volume fractions ( ) above 0.95 tend to be inherently unstable, as the large capillary pressure between the foam bubbles accelerates coalescence. Herein, we demonstrate that these ultradry C/W foams may be stabilized with viscoelastic aqueous phases formed with a single surfactant at a concentration of only 1% (w/v). The C/W foams are stable in DI water and over a wide range of salinity. Moreover, they are stable even for temperatures up to 120oC and with up to 0.99. The properties of the foams and the aqueous viscoelastic solutions are examined for a series of surfactants. For our surfactants, the relaxation times from complex rheology indicate the formation of viscoelastic wormlike micelles over a wide range in salinity and pH, given the high surfactant packing fraction. At 90 oC and above 0.9 , the foam morphology was composed of ~ 35 m diameter bubbles with a polyhedral texture. The apparent viscosities of these ultradry foams reached more than 120 cP with stabilities more than 30-fold over those for foams formed with the non-viscoelastic surfactants. The highly viscous lamellae remain thicker, due to the slower drainage rate, therefore resulting in slower Ostwald ripening relative to foams with non-viscoelastic lamellae. This is shown by measuring the change in bubble diameter with time via optical microscopy. Ultradry C/W foam at elevated temperatures could potentially be useful for CO2 sequestration, improved oil recovery or for hydraulic fracturing with minimal use of water to reduce the requirements for wastewater disposal.

475500 CFD Simulations and Experiments in Multiphase Flow for Predicting Solid Particle Erosion: Challenges and Success Stories (Invited Talk)

Abstract:

Solid particle erosion is a major problem in many industrial applications where solids are entrained in gas and/or liquid flows. For example, erosion of production equipment, well tubing and fittings is a major operating problem that costs the petroleum industry millions of dollars each year. Entrained sand particles in the oil/gas production fluid impinge on the inner surfaces of the pipes, fittings, and valves that result in solid particle erosion. Even in situations when sand control means are utilized such as gravel packing and sand screens, small sand particles can plug sand screens promoting higher flow velocities through other portions of sand screens causing failure of sand screens and allowing sand production. Erosion can cause severe damage to the piping and equipment wall, resulting in loss of equipment and production downtime.

Prediction of erosion in multiphase pipelines is a complex problem due to the lack of understanding of solid particle distribution in the liquid and gas phases and their corresponding velocities. The particle impact velocity is affected by the pipe geometry, carrying fluid properties and velocity, flow pattern, particle size and distribution in the flow. Among different multiphase flow patterns in horizontal and vertical flows, severe erosion damage can occur in annular and slug flows with high gas velocities and low liquid velocities. Although there is a lack of accurate mechanistic models to predict solid particle erosion, there is a need to develop engineering prediction models for multiphase flows.

It is well known that solid particle erosion rates is a strong function of the impacting velocity of particles and also the mass of impacting particles or the particle rate on a certain location. Predicting solid particle erosion in multiphase flow is a complex task due to existence of different flow patterns. The existence of different flow patterns and sand and liquid holdup in vertical and horizontal pipes means that a unique erosion model has to be developed for each flow regime if the model has to account for the number and velocity of impacting particles. In current efforts at the Erosion/Corrosion Research Center, local void fraction measurements by Wire Mesh Sensors (WMS) and Computational Fluid Dynamics (CFD) simulations of multiphase flows are employed to aid in predicting gas-liquid-sand velocities in multiphase flows. For the first time, CFD is used to compute erosion rates in multiphase flow. Local fluid velocities in multiphase flow are used to as a basis to determine representative particle impact velocities. Also based on data and simulations representing sand holdup in several flow regimes, the mass of impacting particles are estimated. Erosion experiments are also conducted on elbows in a 3-inch and 4-inch large scale multiphase flow loop with gas, liquid and sand flowing in vertical and horizontal test sections. Based on the experimental data and simulations for different flow regimes including slug, wet gas and annular flow a mechanistic model is being developed to predict solid particle erosion rates in multiphase flow.

471008 Development of a Mitigation Strategy for Corrosion-Induced Asphaltene Deposition

Abstract:

The objective of this work is to provide insight into the detrimental effect of iron ions, which are produced as a result of corrosion, on the asphaltene precipitation and deposition tendency during asphaltene production. Furthermore, we aim to provide means to control these adverse effects by using chelating agents.

In a series of experiments, the model oils and actual oils were put in contact with aqueous solutions of iron ions to study the effect of these species on the stability of asphaltenes using the indirect method. Also, a new experimental setup, consisting of a multi-section Teflon column packed with carbon steel spheres was built to quantify the deposition of asphaltenes under different operating conditions. This device was used to study the effect of the aqueous iron solutions on the asphaltene deposition tendency.

It was found that the solubility of iron species is relatively high in the oil phase, compared to other cations, and that unlike other metals iron has a significant destabilizing effect on all the asphaltene systems that were tested. Furthermore, the multiphase systems that were analyzed in the column that is packed with carbon steel spheres, showed increasing asphaltene deposition as the concentration of iron in the aqueous phase increased. Different chelating agents were tried to reduce the detrimental effect of iron and some of these agents were successful. A systematic study was conducted to analyze the effect of different variables on the iron-induced asphaltene precipitation and deposition. pH, ionic strength and emulsified water do not seem to affect the asphaltene precipitation tendency as much as the iron concentration. The effect of roughness caused by the corrosion of the steel spheres is quantified on the increasing tendency of asphaltene deposition.

In this work we account for the corrosion of pipelines and asphaltene deposition and the interrelation between them.

According to the evidence that we have found, the corrosion of pipelines might be an aggravating cause for asphaltene deposition in wellbore and surface facilities. By addressing the corrosion problem, asphaltene deposition might actually subside.

471948 A Data-Mining Framework for Uncertainty Analysis in Pipeline Erosion Modeling

Abstract:

A data-mining framework for uncertainty analysis in pipeline erosion modeling

Wei Dai, Selen Cremaschi[a]

[a] Department of Chemical Engineering, Auburn University, AL 36849

selen-cremaschi@auburn.edu

Abstract

In many industrial operations, solid particles move at high speeds in the fluid system and result in serious wear attack. This type of wear is called erosion. Erosion in pipelines is defined as the material removal from the solid surface due to solid particle impingement. This phenomena, especially in multiphase flow systems, is very complex and depends on many factors including fluid and solid characteristics, the pipeline material properties and the geometry of the flow lines. The safe and efficient operation and design of these pipelines requires reliable estimates of erosion rates.

Given the complexity, most of the modeling work focuses on developing empirical or semi-mechanistic models to predict erosion rates. For example, Oka et al. (2005) developed their erosion model using particle impingement in air with empirical constants based on particle properties and hardness of the target materials. Their model is one of the most commonly cited in the literature. Another semi-mechanistic model called 1-D SPPS (Zhang, 2007), which is widely used for predicting erosion rates by the oil and gas industry, was developed with several empirically estimated parameters, like the sharpness factor of particles, Brinell hardness and the empirical constants in the impact angle function. These empirical parameters are calculated using experimental observations. However, the experimental data used in these calculations and also for model validation and uncertainty quantification are, for the most part, collected in small pipe diameters (from 2 to 4 inches). These small pipe sizes do not coincide with the field conditions, where the pipe diameters generally exceed 8 inches. Hence, the predictions of erosion models are routinely extrapolated to conditions where experimental data or even operating experience is not available, and the estimation of erosion-rate prediction uncertainty becomes crucial especially for systems too-costly to fail.The quantification of this uncertainty is especially important during the design phase for subsea applications, as erosion rate allowance, which is set using the erosion rate predictions and its uncertainty, directly impacts the integrity of the facility.

The uncertainty in model predictions can stem from three sources in general: (1) uncertainty in experimental measurements of input conditions, (2) model form uncertainty (i.e., incomplete presentation of the actual system due to lack of knowledge or imprecise experimental observations), and (3) model parameter uncertainty. The experimental data uncertainty usually consists of measurement errors due to both instrumental and human errors. The reliability of the models largely depends on the ability of model form to capture the details of erosion process in enough granularity. The uncertainty in the model parameters results from an inability to accurately quantify the parameters of a model (Shrestha, 2009).

In this talk, a systematic framework is introduced to quantify erosion-rate prediction uncertainty for operating conditions where experimental data are not available, and for a set of newly-collected experimental data points. The framework incorporates the impacts of model form and parameter uncertainties to estimate prediction uncertainties, and combines data clustering and Gaussian Process Modeling approaches with Monte Carlo Simulation.

For estimating erosion-rate prediction uncertainty, we compiled an experimental database of erosion rate measurements from literature. The database contains 586 data points in single or multiphase carrier flows. Eighty percent of the data in the database are collected for gas dominated flows (i.e., gas only, annular, mist and churn flow). The experimental database covers a wide range of input conditions resulting in significantly different erosion rate measurements. The dataset encompasses data collected from six different flow regimes, with wide-range of material properties and production characteristics.

The data clustering is used to capture the similar characteristics of operating conditions and to identify internal data structures present within the database. Among the data clustering approaches available in the literature, k-prototype (Cheung, 2013) is selected as the most appropriate for our dataset due to the existence of categorical variables. It calculates the similarity based on both categorical attributes and numerical attributes, and classifies the given data points into several clusters such that the similarities between objects in the same group are high while the similarities between objects in different groups are low.

Gaussian Process Modeling (GPM, Rasmussen, 2006) is used to estimate the prediction uncertainty stemming from both model form and model parameter uncertainties. The GPM models erosion-rate model discrepancy, which is defined as the difference between experimental erosion rates and the corresponding erosion rate predictions as a Gaussian random process. This process is presented by mean and covariance functions assuming a multivariate normal distribution. The most likely values of mean and covariance function parameters are determined by Maximum Likelihood Estimation (MLE) using experimental data. Once GPM is trained based on the available data set, a set of hyper-parameters can be obtained and used for future model interpolation or extrapolation analysis (Jiang, 2013). A GPM is built for each cluster identified by the data clustering step.

Finally, the Monte Carlo simulation is used to study the influences of data uncertainty due to limited repetition in the experiments. We previously developed a novel approach to estimate experimental data uncertainties in the absence of repetitive experiments and the kernel density estimations of experimental uncertainty for four different measurement approaches (Dai, 2016). The impact of experimental data uncertainty on GPM predictions is assessed in a Monte Carlo framework where the training of GPM is repeated 1000 times with randomized initializations from the kernel density. After 1000 replications, the distribution of model prediction uncertainty in each cluster is obtained. A box plot is used to show the spread of model prediction uncertainties.

The application of the developed framework is demonstrated on one of the well-known erosion model, 1-Dimensional Sand Production Pipe Saver (1-D SPPS, ECRC), which is used extensively in oil and gas industry for erosion predictions. The data clustering approaches divided the database into seven clusters. In the previous studies, we clustered the data based on flow regimes and built GPMs for these clusters (Dai, 2015). A comparison of data clustering based on the k-prototype approach and flow regimes is given in support of the application of data clustering approach. The mean square error (MSE) and area metric (AM) (Ferson, 2008) of the GPM predictions obtained using a fourth fold cross-validation for both approaches are compared. The smallest MSE and AM are 6.79¡Á10^{-9} and 4.39¡Á10^{-5} based on k-prototype approach and 1.48¡Á10^{-8} and 6.44¡Á10^{-5} based on flow regimes. The results suggest the data clustering based on k-prototype approach where smaller MSE and AM are obtained.

Acknowledgement

This work is supported by the Chevron Energy Technology Company. Discussions and comments from the Haijing Gao, Gene Kouba and Janakiram Hariprasad of Chevron and Brenton McLaury, Siamack Shirazi of E/CRC at the University of Tulsa were highly acknowledged.

<>Reference

Cheung, Y.M. and Jia, H., 2013, Categorical-and-numerical-attribute data clustering based on a unified similarity metric without knowing cluster number. Pattern Recognition, 46, 2228-2238.

Dai, W. and Cremaschi, C., 2015. Quantifying Model Uncertainty in Scarce Data Regions ¨C A Case Study of Particle Erosion in Pipelines. The 12th International Symposium on Process Systems Engineering and 25th European Symposium on Computer Aided Process Engineering, Copenhagen, Denmark

Dai, W., Cremaschi, C., Islam M.A., Nukala, R.T., Subramani, H,J., Kouba, G.E. and Gao, H.J., 2016. Uncertainty analysis of multiphase flow ¨C Case studies from erosion, sand transport, liquid entrainment models. 10th North American Multiphase conference, Banff, Canada

Ferson, S., Oberkampf, W. L., and Ginzburg, L., 2008, Model Validation and Predictive Capability for the Thermal Challenge Problem, Computer Methods in Applied Mechanics and Engineering, Vol. 197, No. 29-32, pp 2408-2430.

Jiang, Z., Chen, W., Fu, Y., and Yang, R., 2013, Reliability-Based Design Optimization with Model Bias and Data Uncertainty, SAE International.

Oka, Y. I., Okamura, K., and Yoshida, T., 2005, Practical estimation of erosion damage caused by solid particle impact: Part 1: Effects of impact parameters on a predictive equation, Wear, 259(1-6), page 95-101.

Norman, C. D. (2013). Correlation of porosity uncertainty to productive reservoir volume Society

of Petroleum Engineers.

Papadopoulos, C. E., & Yeung, H. (2001). Uncertainty estimation and monte carlo simulation method. Flow Measurement and Instrumentation, 12(4), 291¨C298.

Rasmussen, C.E. and Williams, C.K. I., 2006, Gaussian Processes for Machine Learning, The MIT Press.

Shrestha, D.L., 2009, Uncertainty Analysis in Rainfall-Runoff Modelling: Application of Machine Learning Techniques, PhD. Dissertation, UNESCO-IHE, the Netherlands

Zhang, Y., Reuterfors, E.P., McLaury, B.S., Shirazi, S.A., and Rybicki, E.F., 2007, Comparison of Computed and Measured Particle Velocities and Erosion in Water and Air Flows, Wear, 263.

465042 Impact of Solid Particles and Their Wettability on the Rheological Behavior of Hydrates

Abstract:

Flow assurance is the most important technical issue facing offshore energy development. Hydrate formation is considered as one of the most important flow assurance problems due to subsequent blockages that can form in these systems. In an oil-dominated system, emulsion stability is an important criterion for hydrate risk management. The extent to which the emulsion droplets remain segregated during hydrate formation/dissociation determines the level of hydrate plug formation. For water-in-oil emulsions stabilized using solid particles, the type of emulsions and their stability are usually influenced by the wettability of solid particles that adsorb on the oil-water interface. Thus, investigating the effect of solid particles and their wettability on hydrate formation would provide useful insights on the hydrate formation process in these systems. In addition to hydrates, scale formation is an important flow assurance problem. Scales are caused by the deposition of mineral salts that are sparingly soluble in water, and they have a significant impact on production.

In the current study, cyclopentane hydrates are studied in model oil systems using solid particles of different wettability to elucidate the effect of particle wettability on hydrate formation and the subsequent rheological behavior. In addition, the effect of scales, such as calcium carbonate and calcium sulphate, on hydrate formation will be investigated. A DHR-3 stress controlled rheometer whose temperature can be controlled between -20°C to 150°C is used to examine the rheological behavior of hydrate forming emulsions. A helical ribbon geometry is used in this work to quantify the rheological behavior of hydrates. The primary objectives of this work are to investigate the effect of solid particles and their wettability, in addition to the presence of scales, on hydrate formation and the subsequent rheological behavior.

465172 Hydrate Formation in Waxy Oil Systems in the Presence of Surfactants and Solid Stabilizers

Abstract:

Flow assurance is a critical aspect of deep-water production and subsea pipeline operations due to the associated operating and maintenance costs involved in managing and mitigating pipeline blockages due to hydrates and wax. Although considerable research has been conducted on each of these systems independently, there is a lack of fundamental understanding on the interfacial properties of hydrates in waxy oil systems. This lack of information motivated the current experimental investigations regarding hydrate formation in waxy oil systems using visualization and micromechanical force (MMF) techniques. For this work, an Olympus BX53 polarized optical microscope equipped with a shear cell and temperature control (-50°C to 450°C) stage is used for investigating hydrate formation in either a surfactant or solid particle laden waxy oil system. The wax concentration is varied, and its effect on hydrate formation and morphology is evaluated using single water drop hydrate experiments. In addition, a micromechanical cohesion force (MMF) method is used for evaluating the hydrate interfacial behavior in these systems. In addition to measuring the cohesion force, this technique also provides insight into the effect of wax and surfactant/solid particles on the hydrate film growth rate. Both experimental techniques will provide new insights to better understand hydrate formation mechanisms and interfacial properties in the presence of wax and stabilizers.

461386 Modeling of Droplet Coalescence and Breakage in Oil-Water Pipeline Using One-Group Interfacial Area Transport

Abstract:

Water fallout in oil transport pipelines has significant impact on corrosion inhibition management for the oil and gas industry. To apply the corrosion inhibitors where and when it is needed is a delicate balance between how water and oil phases interact and the inhibitor's dissolution in each phase. As a precursor, it is extremely important to understand the evolution of water dispersion in oil during various phases of the production cycle.

In this study water is considered as the dispersed phase and its transport described by the one-group interfacial area transport model. The droplet size follows the evolution of interfacial area concentration by considering the impact of turbulent shearing and droplet collision. The one-group model by Wu *et al.* (1997) is adopted as the baseline and modifications are applied to their coalescence and breakage formulations following the work of Abid and Chesters (1994).

The original model of Wu *et al.* (1997) was proposed for bubbly flows, and while their geometrical arguments for droplet collision and breakage probability remain valid; their efficiency formulations can be significantly different. The mechanisms of coalescence drainage and breakage efficiency were found to be highly dependent on droplet size, surface tension, viscosity, flow regime and volume fraction (Abid and Chesters, 1994). The modified formulation was applied to model the bench-scale (2.5" pipe) experiment of Simmons and Azzopardi (2001) and scaled up to predict the full scale (16" pipe) measurements of Fairuzov *et al.* (2000). The results were found to match very well with the full-scale data up to 20% water volume fraction. The model was implemented using commercial computational fluid dynamics software through user accessible subroutines.

References

Abid, S and A. K. Chesters, "The drainage and rupture of partially-mobile files between colliding drops at constant approaching velocity," Int. J. Multiphase Flow, 20, pp. 613-629, 1994.

Fairuzov, Y. V., P. Arenas-Medina, J. Verdejo-Fierro and R. Gonzales-islas, "Flow pattern transitions in horizontal pipelines carrying oil-water mixtures: full scale experiments," J. Energ. Res. Tech., pp. 169-176, 2000.

Simmons, M. J. H., B. J. Azzopardi, "Drop size distributions in dispersed liquid-liquid pipe flow," Int. J. Multiphase Flow, **27**, pp. 843-859, 2001.

Wu, Q, S. Kim and M. Ishii, "One-group interfacial area transport in vertical bubbly flow," Int. J. Heat Mass Transfer, **41**, pp. 1103-1112, 1997.

474837 Flowloop Investigations Towards Gas Hydrate Slurry Transportability

Abstract:

Gas hydrate blockages are often considered the major flow assurance problem during oil and gas production and transportation, mainly because of the fast timescales at which they can form, compared to other flowline solids. Understanding the gas hydrate slurry formation processes and flow properties are important to developing advanced strategies to manage gas hydrate formation in subsea flowlines, as well as other technological applications of gas hydrates. In this paper we present the key lessons learned from flowloop investigations of gas hydrate mitigation strategies in partially dispersed systems. These systems are often considered the most challenging when assessing the risks due to hydrate plug formation. The mitigation strategy of the addition of an anti-agglomerant on partially dispersed systems, based on quaternary ammonium salts, is used to facilitate hydrate slurry flow over a range of conditions, including different water content and mixture velocity. Continuous pumping as well as transient conditions of shut-in (several hours without pumping) and restart (resuming pumping) were applied. The latter transient operations are often very challenging for sustained flow assurance in subsea flowlines. The results from these tests indicate that an anti-agglomerant could be utilized to maintain hydrate slurry transport at low to mid water contents, but high pressure drops can be encountered for higher water content systems.

474838 Hydrate Blockage Risk Assessment in Gas-Dominant Pipelines

Abstract:

Hydrate formation in subsea gas pipelines is a critical design and operating risk, where the complete hydrate avoidance via thermodynamic inhibition can significantly increase the value of offshore production systems. A reduction in the injection rate of hydrate inhibitors, or the use of hydrate management approaches including kinetic inhibition, relies on understanding a comprehensive mechanism for hydrate blockage formation. The development of this conceptual mechanism for gas-dominant pipelines remains an outstanding objective, which has recently been advanced through the collection of laboratory- and flowloop-scale experimental data. This presentation will introduce a new blockage mechanism based on single-pass gas-dominant flowloop experiments collected in Perth, Western Australia and available data from industrial field trials. The new gas-dominant mechanism for hydrate blockage highlights the importance of liquid water entrainment in the gas phase, where aerosolized water droplets can convert to hydrate particles at kinetic rates. The subsequent sedimentation of these particles then enables severe wall build-up and sloughing events, which have been visually confirmed in flowloop experiments and field trials. This new conceptual mechanism suggests that the surface area of water in the gas pipeline is a critical operating parameter, with small diameter pipe segments being at particularly high risk of blockage due to the enhanced liquid droplet entrainment associated with increased gas velocities, and/or due to gas bubble entrainment in liquid collection associated with the bathymetry.

466209 SNG Technology Via Clathrate Hydrates for Large Scale Storage of Natural Gas

Abstract:

Increasing global demand and widespread applications of natural gas (NG) necessitate the development of more effective and efficient methods for NG storage and transportation. Conventional methods of NG storage are either energy intensive - LNG (Liquefied Natural Gas) or pose high safety risk as in CNG (Compressed Natural Gas). SNG (Solidified natural gas) via clathrate hydrates is advantageous due to its characteristic advantages that include high volumetric capacity, non-explosive nature, environmentally benign and being relatively less energy intensive compared to conventional methods of NG storage. However, a significant challenge that impedes the commercial development of SNG technology is the slow rate of hydrate formation. For the first time, we report an increased rate of mixed methane hydrate formation in a simple unstirred tank reactor configuration (UTR) at moderate temperature and pressure conditions employing tetrahydrofuran (THF) as a promoter. THF was observed to function as a thermodynamic as well as an excellent kinetic promoter for hydrate formation at 283.2 K and moderate pressures (3.0-7.2 MPa). Substantial methane gas uptake of 3.52 (±0.13) kmol/m^3 of water is possible even at reduced experimental pressure of 3.0 MPa and 283.2 K in 2 h employing the unstirred reactor configuration. At higher temperature of 293.2 K, similar methane gas uptake and formation kinetics (as that observed at 283.2 K) is possible with the addition of just 100 ppm sodium dodecyl sulphate (SDS) surfactant due to the synergism between THF and SDS. Further, we demonstrate multi-scale experimental validation of our method to a volumetric sample scale-up factor of 120 and internal reactor diameter scale-up factor of 10. It is noted that substantial decrease in the severity of the storage conditions can be achieved by the use of THF as a promoter. SNG technology has a potential for a large scale commercial deployment.

471344 A Water-Immiscible Heavy Guest Molecule for Fast Methane Hydrate Formation

Abstract:

Clathrate hydrates have an ice-like appearance but a three dimensional structure having hydrogen bonded water cavities in which small guest molecules such as methane, hydrogen, carbon dioxide and light organic molecules are encaged. They can play an important role in CO_2 separation and also be a safe and economical medium for natural gas storage. But to apply gas hydrate techniques to industrial processes, we have to overcome the two barriers: (1) Gas hydrates are usually formed at the interface between the bulk gas and water phases. Once the solid hydrates cover the interface, gas hydrate formation decreases significantly because the hydrate layer becomes a barrier for mass transfer, and thus hydrate conversion is very low. (2) The gas hydrate formation kinetics is extremely unfavorable. This talk addresses the circumvention of the mass transfer and kinetic limits by introducing a water-immiscible heavy guest molecule of idomethane to enhance methane hydrate formation. We will demonstrate the key idea for maintaining liquid-liquid interfaces available for continuous methane gas access to the hydrate growth front and achieving a complete conversion to methane-idomethane binary hydrates within 2 minutes in an unstirred reactor system.

470972 Molecular Dynamics Studies of Structure II Hydrate Nucleation Using Advanced Sampling Techniques

Abstract:

Clathrate hydrates are solid, non-stoichiometric combinations of guest molecules entrapped in a solid water lattice. The guest molecules are generally light gases, such as methane or carbon dioxide. Clathrate hydrates are potentially an important energy resource; some estimates suggest there is more energy trapped in hydrate form than in all other fossil fuel sources on earth. Hydrates also represent a major flow assurance hazard in the petroleum industry. Currently active research on applications of clathrate hydrates ranges from energy and flow assurance to gas separations and transportation. Many such applications of clathrate hydrates require a firm understanding of the mechanism of hydrate formation.

The formation of a new hydrate phase occurs from a supersaturated liquid solution, through a process called nucleation. In nucleation, the formation of the new phase must overcome an activation barrier. The nucleation process is inherently a molecular-level phenomenon, occurring on length- and time-scales that are difficult to probe with experimental techniques. The length and time-scales of hydrate nucleation appear well-suited to be studied through molecular simulation. However, in most cases the associated activation barrier is large. Therefore, the nucleation is a rare event – the waiting time between nucleation events is orders-of-magnitude longer than the nucleation events themselves. Correspondingly, in straightforward simulations most of the computational power is expended waiting for the nucleation event to happen, rather than simulating the nucleation events of interest. In order to study hydrate nucleation with molecular dynamics simulations, we use an advanced sampling technique known as forward flux sampling (FFS).

We study the nucleation of sII clathrate hydrates from a supersaturated liquid solution with FFS. The water and guest molecules are described by a coarse-grained model which uses monatomic water and was developed by Jacobson et al.[1] In this presentation, we report results from large-scale FFS calculations comprising over 190,000 individual molecular dynamics trajectories. A committor probability analysis is used in conjunction with FFS to elucidate the nucleation mechanism. We estimate the reaction coordinate with a least squares estimation technique developed for FFS by Borrero and Escobedo.[2] Elements of the labile cluster hypothesis, local structuring hypothesis, and blob-mechanism are observed at different stages of the nucleation. We discuss the ability of several proposed order parameters to capture nucleation of sII hydrates. Our results highlight the various factors at play in clathrate hydrate nucleation.

(1) Jacobson, L. C.; Hujo, W.; Molinero, V. ***J. Phys. Chem. B* 2010**, ***114*** (43), 13796–13807.

(2) Borrero, E. E.; Escobedo, F. A. ***J. Chem. Phys.* 2007**, ***127*** (16), 164101.

467021 Estimation of Direct Transition Mechanism for Molecular Diffusion in Type *I* Gas Hydrates Using Density Functional Theory

Abstract:

Remarkable research efforts have been performed concerning the study of hydrates and organic clathrates using different theoretical approaches, such as molecular equations of state, molecular simulations including both Molecular Dynamics and Monte Carlo techniques, and different ab initio Quantum Mechanics approaches[1-3]. In this work, CO_2 and CH_4 type I hydrates have been studied using electronic Density Functional Theory (DFT)[4]and the Quantum Theory of Atoms in Molecules (QTAIM). Several mechanisms for transport of gases inside hydrates have been proposed so far in literature. Previous results have pointed out the apparent imposibility of some guests passing directly through faces connecting adjacent cages without destroying the water structure. Both types of cells included in the structure of type I hydrate were modeled as isolated double semi-flexible atomic systems. Interaction potentials of guest molecules with the enclathrating cell, when moving between neighbour cells were calculated using B3LYP/6-311+g(d,p) DFT approximation. Our calculations show that direct transitions are feasible through hexagonal and pentagonal faces without compromising the overall structure integrity in opposition to other results previously reported in literature. This stability has been explored using the QTAIM theory and reveals that even in the case that some bond may break during the transition, all of them are recovered, because the face distorsion is absorbed locally by the hydrogen bond network. The validity of the theory level selected has been stated, and the high anisotropy of the guest-cell interaction potential for the molecules analysed is shown, which may be considered in the formulation of hydrate thermodynamic models as equations of state, and also for the description of transport properties.

References:

[1] A. K. Sum, C. A. Koh and E. D. Sloan, Ind. Eng. Chem. Res, 48, 7457, 2009.

[2] Á. Vidal-Vidal, M. Pérez-Rodríguez, J.-P. Torré and M. M. Piñeiro, Phys. Chem. Chem. Phys., 17, 6963, 2015.

[3] Á. Vidal-Vidal, M. Pérez-Rodríguez and M. M. Piñeiro, RSC Advances, 6, 1966, 2016.

[3] Gaussian 09 Revision D.01, Gaussian Inc. Wallingford CT, 2009.

Acknowledgements

The authors acknowledge CESGA (www.cesga.es) in Santiago de Compostela, Spain, for providing access to computing facilities, and Ministerio de Economía y Competitividad (grant n. FIS2015-68910-P), also in Spain, for financial support.

473336 Measurements and Modeling of Gas Hydrate Formation from High Salinity Systems

Abstract:

In this study, measurements and modeling of gas hydrate phase equilibrium have been performed for aqueous solutions with high salt concentration (near-saturated and saturated solutions) to comprehensively study the interplay of hydrate formation and salt precipitation. As no data exist for hydrate phase equilibria, including the triple point for CH_4 [g] – Solution [l] – Hydrate [s] and quadruple one for CH_4 [g] – NaCl [s] – Solution [l] – Hydrate [s], measurements have been performed with undersaturated and saturated solutions of NaCl, $CaCl_2$, KCl, and $MgCl_2$, as single salts and mixtures, using an isochoric method for pressures up to 206 MPa. For the modeling part, we have developed a simple correlation to determine the hydrate suppression temperature as a function of the salt concentration. This correlation can predict the hydrate equilibrium conditions to within about 1 ^{o}C. Moreover, the approach used for the correlation can be readily extended to other salt systems.

461436 Performance of Biodegradable Kinetic Inhibitors on Natural Gas Hydrate Formation

Abstract:

Clathrate hydrates of natural gas are solid crystalline inclusions that form when suitably sized hydrocarbon components into a lattice structures formed by water molecules at adequately high pressures and low temperatures. The formation of gas hydrates is particularly critical in subsea production, transport, and processing facilities where conditions are more suitable. Several alternative strategies exist for managing hydrate formation and growth whereas use of low dosage kinetic inhibitors is one method that has gained attention over the recent two decades as the required active concentration is below 1.0 wt. % . Although kinetic hydrate inhibitors require much lower dosages than thermodynamic inhibitors their use may be restricted due to poor biodegradability. The purpose of this study is to examine and compare the performance of two commercially available synthetic kinetic hydrate inhibitors within the biodegradable class, tested through well-established screening methods.

466837 Towards a Consistent Thermodynamic Theory for Gas-Hydrates

Abstract:

One of the main criticism of van der Waals-Platteeuw theory, as currently applied for gas hydrates, is the large number of parameters that are required to predict the phase equilibria. These parameters are obtained through regression with experimental data on gas hydrate phase equilibria. As a result, these adjustable carry little physical meaning[1]. In this work, we demonstrate a method to substantially reduce the number of fitted parameters to two per guest molecule while maintaining sufficient accuracy in the prediction of the theromodynamic properties of gas hydrates. This has been possible due to availability of accurate cage occupancy data[2]; development of accurate molecular models for water[3] and guest molecules; and development of methods to rigorously compute the partition function of gas hydrates[4]. We demonstrate our method on gas hydrates formed from methane, ethane and carbon dioxide as well as from their binary mixtures.

References

1. A. A. Chialvo, M. Houssa and P. T. Cummings, *J. Phys. Chem. B*, **106**, 442-451, (2002)

2. J. Qin and W. F. Kuhs, *AIChE Journal*, **59**, 2155-2167, (2013)

3. C. Vega and J. L. F. Abascal, *PCCP*, **13**, 19663-19688, (2011)

4. S. Ravipati and S. N. Punnathanam, *J. Phys. Chem. C*, **119**, 12365-12377, (2015)

471365 Exploring the Free Energy Landscape of Gas Hydrate Nucleation By Forward Flux Sampling

Abstract:

Although the importance of gas hydrate has been widely recognized in diverse areas such as energy recovery, flow assurance, global climate change, and gas storage/transportation, a quantitate knowledge of hydrate nucleation mechanism is still missing. The impediment to acquiring this understanding lies mainly in its stochastic nature and ultra-fine scales, making it challenging to explore hydrate nucleation experimentally. Molecular simulation, which is an ideal tool for unveiling process at the molecular level, can only access a very limited volume of the system, thus requiring a long induction time to observe a spontaneous nucleation event under a realistic thermodynamic condition.

Here we present an advanced computational strategy based on forward flux sampling (FFS)[1] to overcome the limit of the direct molecular simulation. Through this strategy, we pursue a quantitative, "first principle" description on the key factors contributing to the crystallization of hydrate. To facilitate the application of FFS in studying methane hydrate formation, we developed an effective order parameter based on the topological analysis of the tetrahedral network[2]. The order parameter capitalizes the signature of hydrate structure, i.e., polyhedral cages, and is capable of efficiently distinguishing hydrate from ice and liquid water while allowing the formation of different hydrate phases, i.e., sI, sII, and amorphous. The quality of the developed order parameter was subsequently verified by the rigorous pB histogram analysis, i.e., committor probability analysis, which indeed showed that the order parameter describes well the reaction coordinates of hydrate nucleation.

Employing the developed method, we explicitly explore hydrate nucleation on its nucleation rate, free energy landscape, and structural evolution, for different guest species, and under conditions where spontaneous hydrate nucleation becomes too slow to occur in direct simulation.

First, analyzing the obtained large ensemble of hydrate nucleation trajectories, we show hydrate formation, on average, is facilitated by a gradual transition from amorphous to crystalline structure, which provides the direct support to the two-step nucleation mechanism of gas hydrate. Importantly, the analysis also reveals the diversity of hydrate nucleation behaviors. In particular, there also exist nucleation pathways where hydrate crystallizes directly into crystalline phase, without necessarily going through the amorphous stage. The diversity of hydrate nucleation pathways highlights the complexity of hydrate nucleation, and calls for further study for hydrate nucleation mechanism under various conditions.

Second, we obtained the free energy landscape along the nucleation pathway through combining the FFS and backward fluxing sampling (BFS)[3]. Since the stationary distribution as a function of the order parameter is calculated independent of nucleation theory, the obtained free energy profile can be used to test the applicability of the classical nucleation theory (CNT). Interestingly, the calculated free energy profile for homogeneous ice nucleation was found to fit perfectly against CNT, a result supporting our previous studies[4,5] demonstrating the classical nature of ice nucleation. In comparison, the calculated free energy profile for hydrate nucleation shows noticeable deviation from the CNT, particularly at small cluster sizes. The difference may not be surprising, and can be attributed to the non-classical nucleation pathways of hydrate and the irregular shape of hydrate nucleus. Nevertheless, the differences aside, the CNT was found to describe the free energy landscape of hydrate nucleation reasonably well, despite its non-classical nature.

Acknowledgement:

The work is supported by NSF through award CBET-1264438.

References:

(1) Allen, R. J.; Frenkel, D.; Wolde, Ten, P. R. Simulating Rare Events in Equilibrium or Nonequilibrium Stochastic Systems. ***J Chem Phys* 2006**, ***124*** (2), 024102.

(2) Bi, Y.; Li, T. Probing Methane Hydrate Nucleation Through the Forward Flux Sampling Method. ***J Phys Chem B* 2014**, ***118*** (47), 13324–13332.

(3) Valeriani, C.; Allen, R. J.; Morelli, M. J.; Frenkel, D.; Rein ten Wolde, P. Computing Stationary Distributions in Equilibrium and Nonequilibrium Systems with Forward Flux Sampling. ***J Chem Phys* 2007**, ***127*** (11), 114109.

(4) Li, T.; Donadio, D.; Russo, G.; Galli, G. Homogeneous Ice Nucleation From Supercooled Water. ***Phys Chem Chem Phys* 2011**, ***13*** (44), 19807–19813.

(5) Cabriolu, R.; Li, T. Ice Nucleation on Carbon Surface Supports the Classical Theory for Heterogeneous Nucleation. ***Physical Review E* 2015**, ***91***, 052402.

455212 Experimental Study of Retention Mechanism of Two Different Types of Scale Inhibitors in Carbonate Formations

Abstract:

When oil reservoirs produce water, then one of the potential production problems is the formation and deposition of mineral scales such as barium sulphate and calcium carbonate. One of the most effective measures for preventing the formation of mineral scale in the oilfield is through the application of chemical scale inhibitors (SI). These inhibitors are usually applied by injecting them into the near-well reservoir formation in a procedure referred to as a "squeeze" treatment. Once these SI chemicals are placed into the formation, they may interact with the formation minerals through mechanisms, such as adsorption or precipitation. It is these interactions that affect the SI retention within the reservoir and the subsequent release of SI in the squeeze treatment.

Since over 50% of the world's oil production takes place from carbonate formations, especially in the Middle East, understanding the chemistry behind the mode of retention of scale inhibitors within carbonate formations is an important issue. Our experimental results show that different adsorption/precipitation regimes are clearly demonstrated in carbonate mineral substrates for the scale inhibitors, DETPMP and VS-Co. Precipitation is more dominant for DETPMP/carbonate retention while pure adsorption is more dominant for VS-Co (but both adsorption and precipitation are observed for both of these SIs). The rather different behaviour of DETPMP and VS-Co may be ascribed to different functional groups having different strengths of SI – Ca binding. Coupled adsorption/precipitation experiments were carried out using two different size fractions of Moroccan calcite (315-500 µm and 125-250 µm) to study the effect of particle size on apparent adsorption. The reason that particle size was investigated was that the SI/Ca complex which precipitates may form a "skin" on the surface, which could affect further SI-Ca interaction, as a "surface poisoning". In fact, our results using both ESEM/EDAX and direct particle size analysis (PSA) clearly show that no (or very little) surface deposition or coating around calcite grains occurred for either of these SIs. Essentially, the results for both particle sizes were the same for each of the SIs. The DETPMP retention showed coupled adsorption/precipitation behaviour which was predominantly precipitation (at [DETPMP] > 100ppm) and this was quantitatively almost the same for both calcite particle sizes. Likewise, for VS-Co the results were quantitatively almost the same for both particle sizes and retention was predominantly by adsorption up to [VS-Co] ~ 3000ppm (although precipitation at higher [VS-Co] was observed). Both sets of adsorption/precipitation experimental results for DETPMP and VS-Co on each calcite particle size fraction were modelled using a model developed previously and quantitative matches to the data were obtained. This was carried out by matching the results for one (m/V) ratio* and then predicting the results for the other (m/V) ratio (* m/V is the ratio of mass of mineral substrate to the volume of SI solution in the "apparent adsorption" experiments).

As there is some debate about the common retention mechanism of phosphonate and particularly polymeric scale inhibitors and also the effect of particle size on apparent adsorption of these SIs in carbonate systems, understanding the surface chemistry of the retention mode of scale inhibitors and skin formed by precipitation of a SI/Ca complex around the calcite grains and consequently surface poisoning, by performing static adsorption/ compatibility tests, ESEM/EDAX and Particle Size Analysis, can help production engineers to design improved inhibitor squeeze treatments. Subsequently, longer squeeze lifetimes and improved efficiency of SI deployment in carbonate reservoirs could be achieved.

Keywords: Scale Inhibitors, Squeeze Treatment, Retention, Adsorption, Precipitation

460435 Small-Angle Scattering Investigation of the Fractal Structure of Petroleum Asphaltenes

Abstract:

Asphaltene fractal clusters can be destabilized by changes in the composition, pressure, and/or temperature of a petroleum mixture [1,2]. It is well known that asphaltenes self-assemble into large-scale and thermodynamically stable fractal clusters [1,2]; however, the mechanism of this assembly and why the size terminates on the nanometer length scale is currently unclear. In order to better understand the effect of solvent environment on the fractal structure and interaction of asphaltenes, small angle scattering (SAXS) and ultra-small angle x-ray scattering (USAXS) experiments were performed to identify the fractal cluster's size, shape, and molecular weight in various solvents and mixtures. This talk will first present SAXS results of asphaltenes as a function of concentration in various solvents and anti-solvents generated on beamline 12-ID-C at the Advanced Photon Source at Argonne National Lab. Understanding how solvents and anti-solvents influence the fractal dimension and dissociation of asphaltene clusters is critical in understanding the asphaltene destabilization mechanism. The SAXS experiments were designed to understand the thermodynamics of the asphaltene nanoaggregate and cluster association/dissociation process. USAXS results generated on beamline 9-ID-C will also be presented to explore the asphaltene precipitation process. Because USAXS is able to detect a wider range of scattering vectors (q), both soluble and precipitating asphaltenes with larger length scales can be observed. Overall, this study can provide an improved understanding of the solvent's influence on the structure and association of asphaltenes.

[1] Hoepfner, M. P., Fávero, C. V. B., Haji-Akbari, N., and Fogler, H. S, "The fractal aggregation of asphaltenes," Langmuir, vol. 29, 2013, pp. 8799–808.

[2] Hoepfner, M. P., and Fogler, H. S., "Multi-scale Scattering Investigations of Asphaltene Cluster Breakup", Langmuir, vol. 29, 2013, pp. 15423–15432

460822 A Multicomponent Diffusional Model for Adsorption Dynamics of Asphaltenes at Water/Oil Interfaces

Abstract:

*Corresponding author e-mail: vpauchard@ccny.cuny.edu

Keywords: Asphaltenes, Mixture Effects, Diffusion, Dilatational Rheology

Abstract

Mixture effects were introduced in a previous communication[1] to explain the deviation between single component models of diffusion-controlled adsorption and experimental data for adsorption of asphaltenes at water and organic phase interface. A two-component model coupling Langmuir isotherm, Langmuir equation of state and diffusion kinetics was demonstrated to capture with only four adjustable parameters (2 concentrations and 2 adsorption coefficient) both dynamic interfacial tension over 4 hours and dilatational rheology at various adsorption times (over two frequency decades). Only a few percent of the total asphaltenes solubility class was found to be significantly surface active, consistently with the observation that only a small fraction of them actually contribute to emulsion stability. However due to the diversity of asphaltenes molecular structures (and of their reported interfacial activity), the two-component model still suffers from some inaccuracies. This is particularly apparent at short times when adsorption is expected to be controlled by the bulk of asphaltenes. Increasing the number of pseudo-components would be expected to solve those issues. Unfortunately, due to an extreme mathematical complexity, no analytical model is available for diffusion controlled dilatational rheology with more than two components. This difficulty can be circumvented by numerically solving the set of first principle equations governing relaxation during dilatational rheology experiments (for each component one mass conservation/diffusion equation and one adsorption isotherm equation). A matrix type solving procedure is proposed that proves both extremely simple and robust. This procedure is applied to the well-known dataset from Freer and Radke who concluded to a non-diffusion controlled rheological behaviour. With a unique set of six adjustable parameters (3 concentrations, 3 adsorption coefficients) the mixture model captures with a great level of detail both dynamic interfacial tension over 24 hours and dilatational rheology over 7 frequency decades. Interestingly, the parameter optimization procedure naturally comes out with a total asphaltenes concentration (the sum of the 3 pseudo-components concentration) nearly equal to the nominal concentration of the asphaltenes solution. Some attempts to account for non-planarity of the droplet surface on diffusion kinetics and to model desorption during washout seem promising.

1 Mixture Effects on Adsorption Kinetics of Asphaltenes at the Water/Toluene Interface. Vincent Pauchard and Fang Liu. 2016 AICHE Spring Meeting. April 10-14, 2016. Houston.

2 Rane, J. P., Pauchard, V., Couzis, A., & Banerjee, S. (2013). Interfacial rheology of asphaltenes at oilâ€"water interfaces and interpretation of the equation of state. Langmuir, 29(15), 4750-4759.

3 Freer, E. M., & Radke, C. J. (2004). Relaxation of asphaltenes at the toluene/water interface: Diffusion exchange and surface rearrangement. Journal of Adhesion, 80(6), 481-496.

463674 Comparison of the Vapor-Liquid Equilibria and Solubility Predictions of Various Gas Mixtures Using the Cpa , SAFT and PC-SAFT Â Equations of State

Abstract:

Reliable phase equilibria calculations are vital for the design of several hydrate applications, including flow assurance, and desalination and gas separation technologies using hydrates. The fundamental technical and economic feasibility of these latter processes are dependent on the accurate assessment of phase equilibria behavior of the fluid and hydrate phases. In order to calculate accurate phase equilibria behavior of hydrates, the fluid models need to be corrected for pure component properties and binary vapor-liquid equilibria over a range of temperatures and pressures. Five parameters (a, b and c for the cubic EoS part, and two for the association volume and strength) of the cubic plus association equation of state were optimized for the associating components by simultaneous minimization of absolute errors in saturated liquid densities and vapor pressures with comparison to experimental data and DIPPR (Design Institute of Physical Properties) correlations .Various other equations of state, which include the cubic plus association (CPA-EoS), Soave Redlich Kwong (SRK), Peng Robinson (PR), Statistical Associating Fluid Theory (SAFT), and Perturbed-Chain SAFT (PC-SAFT), have been applied in this work to predict vapor-liquid equilibria of various binary systems (Hydrate Former (HF) + hydrocarbons). SAFT and PC-SAFT equation of state parameters were adopted from [1,2]. Model accuracies were compared with and without using binary interaction parameters for various binary systems containing C_1-nC_3/CO_2/H_2S/N_2/MEG/MeOH/EthOH/H_2 + hydrocarbons (low to medium molecular weight). Furthermore, absolute errors (*AADx*, *AADy*) were calculated and compared for combinations of binary mixtures, including self associating systems. In predicting the vapor part of phase envelope, all equations of state gives comparable accuracy. However, predictions of the liquid part of the phase envelope show significant errors without using BIP for all EoS (SRK, PR, cPA, SAFT and PC-SAFT EOS). Moreover appropriate mixing rules for cPA, SAFT and PC-SAFT equations of state are required to reduce the deviations from experiment when these models are applied to polar molecules. The association energy and volume of interaction are treated with mixing rules developed by [3], to investigate the effect of various mixing rules on model accuracies.

463783 Amino Acids As Environmentally Friendly Low Dosage Hydrate Inhibitors for Offshore Flowassurance

Abstract:

Title : Amino Acids as environmentally friendly low dosage hydrate inhibitors for offshore flow assurance

M. Fahed Qureshi , Mert Atilhan, Tausif Al-Tamash ,

Department of Chemical Engineering, Qatar University

Corresponding Authors: M.Atilhan@qu.edu.qa / MQureshi@qu.edu.qa / Tausif.AlTamash@qu.edu.qa

Abstracts:

Gas hydrates are metastable crystalline compounds that are formed when small gas molecules like ethane and methane gets trapped within the water molecules, at higher pressure and lower temperature conditions. These gas hydrates tend to accumulate in the offshore subsea lines and over the period tend to completely block the pipeline, halting the gas and oil supply to the onshore stations. Therefore, annually the industry spends over 100 million dollars on hydrate prevention procedures. These procedures include mainly the use of chemical thermodynamic hydrate inhibitors (THI) and Kinetic hydrate inhibitors (KHI). The THI like Mono-ethylene glycol (MEG) and Methanol are mainly used in industry. However, these THI are required in bulk quantities, which causes a major disposal issue and also post a major safety threat to nearby operators due to high flammability of Methanol. These commercial THI are also not suitable for environmental, so there is a strong research interest in finding a new class of low dosage KHI and THI inhibitors that are environmentally friendly and effective as commercial THI and KHI.

Amino acids are biological organic compounds containing amine (-NH_2) and carboxylic acid functional groups (-COOH) that have been previously reported to be an environmentally friendly corrosion inhibitors. So the feasibility of using amino acids as a potential KHI and THI have been tested using pure methane gas on a rocking cell assembly within the pressure range of 40-120 bars. The types of amino acids tested include: L-Alanine, L-Phenylalanine, Glycine, Histadine and L-Asparagine.

The experimental results obtained show that the amino acids with lower solubility (0.02g/ml) provided significant KHI effect, but no THI effect. However, the amino acids with higher solubility (0.04g /ml - 0.05g/ml) provided both THI and KHI effect simultaneously. The amino acids with higher solubility were able to shift hydrate vapour liquid equilibrium curve (HLVE) by 0.7 $^{\circ}$ C at both lower (40 bars) and higher (100 bars) pressures. These amino acids were also able to delay methane hydrate crystal formation by up to 1 hour at both lower (40 bars) and higher (100 bars) pressures. Thus, the selected amino acids with higher solubility were found to have a have a potential to act as a KHI and THI both simultaneously.

Acknowledgement

This work was made possible by NPRP grant # 6-330-2-140 and GSRA # 2-1-0603-14012 from the Qatar National Research Fund (a member of Qatar Foundation). The statements made herein are solely the responsibility of the authors.

466019 Inhibited Gas Hydrate Formation at Activated Carbon Particle-Laden Oil-Water Interfaces

Abstract:

This study showed that the formation of cyclopentane (CP)-propane mixed hydrate is inhibited by activated carbon particles in the non-hydrate forming oil phase (isooctane) having a nonionic surfactant (sorbitan monooleate, Span 80). Due to the hydrophobicity, the activated carbon layer is formed at the oil (CP)-water interface. The inhibition effects of activated carbon layer come from preventing the contact between water molecules and hydrate formers. First, the formation of cyclopentane-propane mixed hydrate was confirmed using Raman spectroscopy; the results of Raman spectra showed that cyclopentane and propane mixed hydrate is structure II. Using high pressure micro differential scanning calorimetry (HP micro-DSC), the inhibition efficiency of activated carbon was analyzed by measuring the dissociation enthalpy. The dissociation enthalpy of mixed hydrate was reduced sharply until 0.5 wt% of activated carbon. When the activated carbon is 1.0 wt%, because the oil-water interface is covered entirely by the particle layer, and thus, hydrate growth was not observed over 6 hours. These results indicate that activated carbon particles can be used as a kinetic hydrate inhibitor.

469872 Hydrate Formation in Pickering Emulsions in Flowing Conditions

Abstract:

In the current study, cyclopentane hydrates are studied in model oil systems. Experiments are performed using a 1 inch flow loop. The process tubing is jacketed to facilitate the flow of a temperature control fluid in the annulus between the jacket and the process tube in order to control the temperature. The temperature range of the flow loop can be controlled in the range from -30°C to 125°C using a re-circulating chiller. Fluid is circulated through the loop using a progressive cavity pump that can pump fluid up to 20 gpm at pressures up to 150 psig. An in-flow microscope is utilized to visualize the flow behavior of concentrated solid stabilized emulsions during hydrate formation. Pressure drop measurements will complement the in-flow microscopy measurements in order to provide a comprehensive assessment of hydrate formation and deposition in these systems. The primary objective of this work is to quantify hydrate formation and deposition behavior in solid-stabilized emulsions under flowing conditions.

AUTHOR INDEX

Abedi, Jalal 18
Ahn, Yun-Ho 49
Aichele, Clint P. 43, 44, 62
Akhmetkhanova, Nelya 58
Akkutlu, I. Yucel 20
Alrashdan, Saad 27
Altamash, Tausif 60
Alzobaidi, Shehab 37
Aman, Zachary M. 44, 47
Arca, Serhat 26
Atilhan, Mert 60
Badran, Ismail 23
Baek, Seungjun 49, 61
Baskent, Emel 26
Bbosa, Ben 46
Beck, Griffin 33
Bhagwat, Swanand 62
Bi, Yuanfei 55
Bluck, David 6
Boak, Lorraine. S. 56
Bui, Khoa 20
Campbell, Darin 22
Caniaz, Ramazan O. 26
Caputo, Dominic 29
Cetintas, Refika 26
Chaisoontornyotin, Wattana 16, 57
Charlton, Thomas 47
Chen, Andrew 7, 39
Chen, Chau-Chyun 4
Cho, Hyeyoung 29
Clark, Peter 43, 44
Cremaschi, Selen 40
Da, Chang 37
Dai, Wei 40
Dapena, Jose 46
Daraboina, Nagu 3, 53
Darjani, Shaghayegh 58
Defever, Ryan 50
Deo, Milind 29
El-Qanni, Amjad 23
Elsharkawy, Adel 27
Gerek, Nevin 6
Grewal, Greg 5
Hao, Yifan 4
Harding, Thomas G. 35
Hassanzadeh, Hassan 18, 35
He, Lin 31
He, Peng 21
Hirohama, Seiya 6
Hoepfner, Michael P. 16, 19, 57
Hoopes, Kevin 33
Hu, Yue 52
Huang, Hai 36
Islam, M. R. 4
Jarboe, John 5
Jarrahian, Khosro 56
Johns, Michael L. 47
Johnston, Keith P. 37
Kang, Hyery 49
Kannan, Seetharaman Navaneetha 3
Khan, M. Naveed 46
Khan, Muhammad 59
Kim, Doo-Wook 24
Kirkes, Toni 4
Kocaman, Elif 26
Koh, Carolyn A. 46, 47, 59
Koh, Dong-Yeun 49
Koltveit, Alexander 53
Kostereli, Ziya 26
Koteeswaran, Samyukta 43
Lee, Bo Ram 52
Lee, Chang-Ha 24
Lee, Huen 49
Lee, Jae W. 49, 61
Li, Tianshu 55
Li, Xingang 31
Lin, Helen 8
Linga, Praveen 48
Liu, Fang 58
Maheshwari, Utkarsh 32, 34
Majid, Ahmad Abdul 46
Makinde, Ibukun 30
Manasrah, Abdallah 23
Martin, Tyler 62
Martinez, Alberto 29
May, Eric F. 47
Min, Juwon 61
Morrissy, Shane A. 44
Nassar, Nashaat N. 23
Nikolaou, Michael 30
Oakey, John 17
Olsen, Tim 28
Pauchard, Vincent 58
Pawar, Gorakh 36
Pérez-Rodríguez, Martín 51
Peters, Cor 59
Phatak, Alhad 33
Piñeiro, Manuel M. 51
Poerner, Melissa 33
Punnathanam, Sudeep 54
Qureshi, M. Fahed 60
Raman, Ashwin Yegya 43, 62
Ramos-Pallares, Francisco 8
Sabet, Nasser 18
Sadeghi, Asghar 35
Samson, Ng 16
Sarica, Cem 3
Sarupria, Sapna 50
Satyro, Marco 8
Senra, Michael 5

AUTHOR INDEX

Shaffer, Daniel....6
Shirazi, Siamack....38
Shukla, Utsav....32
Singh, Kaveri....34
Singleton, Mike. A....56
Song, Hua....21
Sorbie, Kenneth. S....1, 56
Sridhar, Garud....33
Srivastava, Vishal....46
Sui, Hong....31
Sum, Amadeu K....52
Sung, Chi-An....39
Tavakkoli, Mohammad....7, 39
Taylor, Shawn D....8
Tran, Vu....37
Trinh, Tuan....6
Tsai, Kuochen....45
Vargas, Francisco....7, 39
Veluswamy, Hari Prakash....48
Venkataramani, Deepika....44, 62
Venkatesan, Rama....3
Verma, Sandeep....33
Verri, Giulia....1
Vessam, Shivanand....54
Vidal-Vidal, Ángel....51
Volk, Michael....46
Von Solms, Nicolas....53
Wang, Junyan....31
Wang, Meng....4
Wang, Shu....22
Warrier, Pramod K. W. Harikumar....59
Watanasiri, Suphat....22
Waturuocha, Amaka....46
Yang, Yuan....19, 57
Yao, Jia....17
Yarranton, Harvey W....8
Yasar, Muzaffer....25
Yasar, Sebnem Hande Ozoren....25